Advancements in Fine Particle Plasmas

Edited by Aamir Shahzad

Published in London, United Kingdom

Advancements in Fine Particle Plasmas
http://dx.doi.org/10.5772/intechopen.111068
Edited by Aamir Shahzad

Contributors
Aamir Shahzad, Fazeelat Hanif, Alina Manzoor, Qurat-Ul Ain Asif, Ines Sarah Medjahdi, Abdel Karim Ferouani, Mohammed Sahlaoui, Mostefa Lemerini, Bheem Singh Jatav, Banashree Saikia, Paramananda Deka

© The Editor(s) and the Author(s) 2024

The rights of the editor(s) and the author(s) have been asserted in accordance with the Copyright, Designs and Patents Act 1988. All rights to the book as a whole are reserved by INTECHOPEN LIMITED. The book as a whole (compilation) cannot be reproduced, distributed or used for commercial or non-commercial purposes without INTECHOPEN LIMITED's written permission. Enquiries concerning the use of the book should be directed to INTECHOPEN LIMITED rights and permissions department (permissions@intechopen.com).

Violations are liable to prosecution under the governing Copyright Law.

Individual chapters of this publication are distributed under the terms of the Creative Commons Attribution 3.0 Unported License which permits commercial use, distribution and reproduction of the individual chapters, provided the original author(s) and source publication are appropriately acknowledged. If so indicated, certain images may not be included under the Creative Commons license. In such cases users will need to obtain permission from the license holder to reproduce the material. More details and guidelines concerning content reuse and adaptation can be found at http://www.intechopen.com/copyright-policy.html.

Notice

Statements and opinions expressed in the chapters are these of the individual contributors and not necessarily those of the editors or publisher. No responsibility is accepted for the accuracy of information contained in the published chapters. The publisher assumes no responsibility for any damage or injury to persons or property arising out of the use of any materials, instructions, methods or ideas contained in the book.

First published in London, United Kingdom, 2024 by IntechOpen
IntechOpen is the global imprint of INTECHOPEN LIMITED, registered in England and Wales, registration number: 11086078, 5 Princes Gate Court, London, SW7 2QJ, United Kingdom

British Library Cataloguing-in-Publication Data
A catalogue record for this book is available from the British Library

Additional hard and PDF copies can be obtained from orders@intechopen.com

Advancements in Fine Particle Plasmas
Edited by Aamir Shahzad
p. cm.
Print ISBN 978-1-83769-790-8
Online ISBN 978-1-83769-789-2
eBook (PDF) ISBN 978-1-83769-791-5

We are IntechOpen,
the world's leading publisher of
Open Access books

Built by scientists, for scientists

6,900+
Open access books available

185,000+
International authors and editors

200M+
Downloads

156
Countries delivered to

Our authors are among the
Top 1%
most cited scientists

12.2%
Contributors from top 500 universities

WEB OF SCIENCE™

Selection of our books indexed in the Book Citation Index
in Web of Science™ Core Collection (BKCI)

Interested in publishing with us?
Contact book.department@intechopen.com

Numbers displayed above are based on latest data collected.
For more information visit www.intechopen.com

For EU product safety concerns:
IN TECH d.o.o., Prolaz Marije Krucifikse Kozulić 3, 51000 Rijeka, Croatia,
info@intechopen.com or visit our website at intechopen.com.

Meet the editor

Aamir Shahzad has more than 19 years of university research and teaching experience at home and abroad. He studied the thermophysical properties of materials and plasmas at graduate and postgraduate levels. He received his post-doctorate and doctoral degrees from Xi'an Jiaotong University (XJTU), China, in 2015 and 2012, respectively. He has proposed novel methods to explore outcomes of complex materials, highlighting his aptitude in computational physics and molecular modeling and simulations. Moreover, Dr. Shahzad is interested in computational physics, complex fluids/plasmas, plasma oncology, and bio and energy materials. Currently, Dr. Shahzad is an associate professor in the Department of Physics at Government College University Faisalabad (GCUF), Pakistan. He is a member of the XJTU ThermoPhysical Society, the GCUF, and the University of Agriculture Faisalabad Physics Societies.

Contents

Preface

Recent research on fine particle plasmas, also known as dusty plasmas, began with theoretical investigations of the origin and presence of fine/dust particles and their observations. This book discusses various advancements in fine particle plasma or dusty plasma research, including new theories, models, algorithms, and the applications of fine particle plasma to the solution of practical plasma science and technology problems. Research on dusty plasmas in fundamental basic sciences, as well as their applications, has increased significantly in different modern technologies. Advancements in Fine Particle Plasmas discusses multidisciplinary theoretical and experimental studies on the fundamental transport behaviors, interaction physics, and reaction chemistry of fine particles and their applications to technologies ranging from plasma lighting sources, plasma refrigeration, and lasers and sensors to materials processing, agriculture and food processing, biotechnology, health care, plasma medicine, antibacterial/antiviral applications, microelectronics fabrication/semiconductor industries, and space and environmental sciences. From an experimental and computational perspective, plasma devices are challenging to operate and model. The problem is inherently multiphysics due to the collective behaviors of fine particle plasma with the surrounding conducting/nonconducting structures. The investigation of Coulomb crystals is also a basic physical problem, for instance, strongly coupled plasmas, phase transition, electric and magnetic effects, and complex phenomena.

This book provides an introduction to and presents current research on fine particle plasmas. Nowadays, the advancements in industrial fabrication engineering using plasma technology and its advanced manufacturing technologies have a significant role in modern innovative applications. The book identifies key challenges, including basic interactions of fine particles with materials, living tissues, fine particle plasmas that contain chaotic and stochastic processes, very high-density plasma and microplasmas, and high-temperature and high pressure plasmas. Advancement in life sciences involves the deployment of discrete technologies and information, and fine particle plasma technology is a key advancement in modern medicine and health care. The use of fine particle plasma in agriculture and medicine is a relatively new area that arose from research on the application of dusty plasma in bioengineering. Fine particle plasma is an innovative and promising novel, multidisciplinary field of research covering plasma physics, life sciences, clinical medicine, food safety and technology, and sustainable food chains.

Dr. Aamir Shahzad
Professor (Associate),
Modeling and Simulation Laboratory,
Department of Physics,
Government College University,
Faisalabad, Pakistan

Chapter 1

Introductory Chapter: Progress of Plasma Physics and Allied Technologies in Daily Life Applications

Aamir Shahzad, Fazeelat Hanif, Alina Manzoor and Qurat-Ul Ain Asif

1. Introduction

"Plasma is a quasineutral gas of charged and neutral particles which shows collective behavior." The neutral gas ionization contains equal numbers of positive and negative charge carriers ($n_i \approx n_e \approx n$). Where n_i is ion density, n_e is electron density, and n is number density. Such plasma is called "quasineutral." Irving Langmuir in 1926, first used the term plasma to explain an inner region of electrical discharge. Plasma has different types depending on many factors such as pressure, temperature, and charges particle density. Commonly used types are nonthermal atmospheric pressure plasma (cold plasma) where $T_e > T_i > T_g$, and hot plasma where $T_e \approx T_i$. The temperatures of electrons, ions, and gas molecules are shown here as T_e, T_i, and T_g, respectively [1]. Cold atmospheric plasma (CAP) is produced by subjecting gases such as helium, argon, nitrogen, oxygen, or air to a high-voltage, high-frequency electrical discharge. In large number of applications, nonthermal plasma is commonly used as they produce a variety of reactive species, such as reactive nitrogen species (RNS), reactive oxygen species (ROS), or both reactive oxygen and nitrogen species (RONS) [2]. Plasma physics is now emerging at a fast pace. Some of its applications are plasma medicine, plasma agriculture, microfabrication, and environmental science which are described below (**Figure 1**).

1.1 Plasma medicine

Plasma medical science is a unique interdisciplinary field that includes studies on plasma science and medical science. Cancer, wound healing, blood coagulation, regenerative medicine, and dental treatment have all benefited from the use of nonthermal atmospheric pressure plasma (NTP) in medicine. It has been extensively studied how plasma interacts with cells and tissues. NTP can be used in two different ways in plasma medicine. The first is direct treatment, which involves bringing plasma into direct contact with the biological material to be treated. In this treatment, all plasma-generated species, both short lived and long lived species come into close contact with the sample and work synergistically. The second method is known as indirect treatment. In this procedure, plasma is used to activate

 IntechOpen

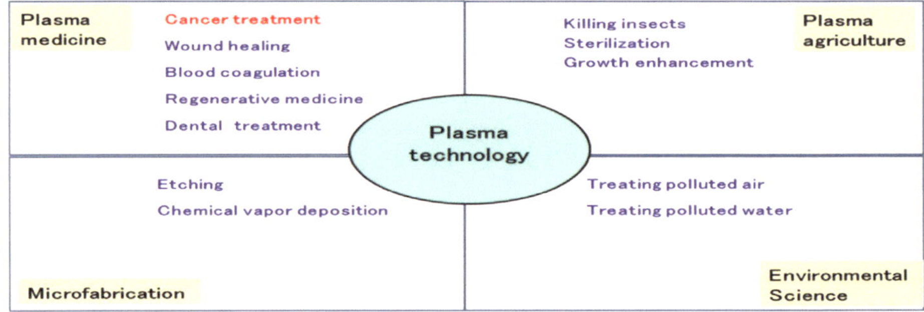

Figure 1.
Wide applications of plasma technology in daily life. There are many applications of plasma such as plasma medicine, plasma agriculture, microfabrication, and environment science [3].

a liquid medium, which is then used for treatment. Only long-lived chemical species are involved in this case [4]. Plasma-assisted immune therapy and plasma-activated medium are two examples of indirect treatments. In plasma cancer treatment, managing intracellular redox equilibrium may be crucial. Animal studies are necessary to evaluate the efficacy and safety of these medicines for potential therapeutic use. The use of plasma diagnostics, such as vacuum ultraviolet absorption spectroscopy (VUVAS), optical emission spectroscopy (OES), two-photon absorption laser-induced fluorescence (TALIF), and laser-induced fluorescence techniques, is crucial for understanding the features of plasma. Electron spin resonance (ESR) and colorimetric assays are two techniques for examining liquids exposed to plasma. Techniques include cell viability assays, immunostaining, live imaging, apoptosis assays, reactive oxygen species (ROS) assays, western blotting, flow cytometry, and real-time polymerase chain reaction (PCR), which can be used to examine cells exposed to plasma. Pen-type plasma sources are important for practical use in medicine. Sakakita and Ikehara have developed a mild-type plasma generator for blood coagulation [3]. Platelet-rich plasma (PRP) is a regenerative medicine. PRP is frequently employed as fillers or in operations for skin rejuvenation in the fields of cosmetic medicine and cosmetology. It is injected at the musculoskeletal and tendon injury site, and early results indicate promising structural and functional benefits. A recognized and accepted practice in dental treatment regimens is PRP therapy. PRP injections into nonhealing or chronic wound sites or their use as a dressing have demonstrated promising therapeutic results in wound healing trials. The capacity of PRP to produce various growth factors, cytokines, and interferons is thought to be the cause of its therapeutic effects. When used in addition to traditional wound care procedures, they encourage fibroblast proliferation, angiogenesis, epithelial cell proliferation, and wound healing [5, 6]. In addition, carbon monoxide (CO) also shows a synergistic effect with plasma. It plays a key role in the suppression of organ graft rejection, cardiopulmonary bypass surgery, and pancreatic cancer [7].

1.2 Plasma agriculture

The research on NTPs used to treat seeds and plants that were done with knowledge gained in plasma medicine served as the starting point and the basis for the new field known as "Plasma Agriculture." Such interdisciplinary study is required due to the population growth-related constant rise in food demand. One such strategy is using NTPs in agriculture, which allows for a significant output improvement without additional water or chemical fertilizer. NTPs are extremely effective and valuable, and they have already demonstrated this in several applications in the food sector, completing the entire food cycle. For a very long time, plasmas in the food processing industry were restricted to cleaning packaging materials or their treatment to increase their wettability, sealability, printability, and barrier qualities [8]. Typically, seed treatment protects it against pathogens that could harm the seed during the early stages of seedling development. The agriculture industry's top concern is the presence of microorganisms (bacteria and fungi) that will impact crops. By employing plasma processing techniques, bacteria can be inactivated while seed germination and crop growth are accelerated. Some plasma processing techniques provide a variety of benefits, including quick processing times and low treatment temperatures. To guarantee speedy and uniform germination, more effective seed treatment solutions should not contain hazardous residues, require minimal energy, have a modest penetration depth to avoid harming cells, and favor lengthy storage period while supporting good seed growth. NTP therapies may be able to meet these criteria. By etching, adding functional groups, and covering with unusual materials, plasma can interact with seeds and change their surface features. Plasma treatment of seeds can improve crop germination and seedling growth by removing microbial layers and changing water intake and other factors [9].

Pesticides are sprayed on crops in agricultural fields and plant protection facilities to shield them against a variety of insects and viruses. Foods including cereals, fruits, vegetables, and meats can become contaminated by fungi like Aspergillus or Penicillium. The environment and the human body are both harmed by leftover agricultural chemicals, such as thiabendazole, imazalil, and ortho-phenylphenol. Additionally, methyl bromide is a powerful and frequently employed pesticide. When used for inactivation, NTP and low-pressure plasmas showed promise as a very effective technique that produces the least amount of harm to crops, foods, seeds, people, and the environment [10, 11]. Further, salt-induced suppression of rice seed germination can be alleviated by using an exogenous CO aqueous solution. CO can enhance seed germination. CO has also been shown to stimulate lateral root development in various seeds [7].

1.3 Microfabrication

The development of plasma science and technology has significantly emphasized plasma-assisted microfabrication, particularly since the 1970s. This growing tendency can be largely attributed to "Moore's Law," which states that the number of transistors in a compact integrated circuit doubles approximately every two years [12]. Low pressure, NTPs have several significant uses in the processing of semiconductors for producing microelectronic devices and integrated circuits. Sputtering, plasma-enhanced chemical vapor deposition (PECVD), plasma oxidation (anodization), and planarization are a few examples of film deposition processes. Film removal procedures include etching, stripping, cleaning, and plasma resist development. Currently, polycrystalline silicon ("poly"), silicon dioxide ("oxide"), and metals (often interconnect materials like aluminum-copper alloys and, increasingly, tungsten) are etched using plasma at essential steps. Amorphous silicon, silicon dioxide, and silicon nitride thin films are frequently deposited using (PECVD) techniques. Related technologies for producing large-area thin film transistors for flat panel displays are of great interest [13].

1.4 Environmental science

Volatile organic compounds (VOCs) are contaminants present in various industrial settings, including paint, coatings, and chemical manufacturing facilities as well as semiconductor manufacturing facilities. Utilizing NTP produced at atmospheric pressure or above is one of these traditional VOC abatement techniques. High energy electrons and low gas temperatures are characteristics of atmospheric pressure nonthermal discharges such as DC or AC corona discharges, microwave plasma, and dielectric barrier discharges (DBDs). After years of research, plasmas are showing promise as a VOC remediation technique. Carbon dioxide and water are the major byproducts of the VOC breakdown if enough oxygen is available. Since almost 150 years ago, ozone has been produced industrially using the DBD technique, which is frequently utilized for NTP production at atmospheric pressure. The actual gas temperature is close to ambient temperature, which is cool enough to provide a noncorrosive processing environment while still supporting a variety of chemical reactions that destroy the pollutant. In contrast to the energetic electrons, whose temperatures range from 10,000 to 100,000 K, active radicals, ionic, and excited and dissociated atomic and molecular species, can start plasma chemical processes. DBDs have a significant advantage over other discharges in that the reaction conditions of plasma processing can be adjusted to maximize the average energy of the electrons by varying the gas pressure (or gas density) or discharge gap width. Plasma technology for air pollution control currently has a low energy efficiency and produces harmful byproducts such as CO, NOx, and nitric acid when processing in air streams, despite its demonstrated ability to destroy waste gas streams [14, 15].

2. Green technology

One of the industrial sectors that contributed significantly to the advancement of human civilization throughout history is the textile industry, the second-largest industry in the world. But textiles and its end products consume a lot of water, complex chemicals, and energy, making it one of the top 10 most polluting world's industries. Application of new and innovative technologies is necessary to address the primary demands of environmental protection from pollution. The large area of plasma science has one such technology. Plasma technology has been used in several applications and is now recognized as a flexible tool for improving industrial processes. NTPs are excellent for modifying the surfaces of heat-sensitive textile polymers. The pretreatment and finishing of textiles are done using plasma because it improves the surface properties of polymers without changing their bulk qualities. In order to decrease energy use and pollutant production, NTP integration into traditional energy-intensive wet-chemical textile manufacturing is becoming increasingly popular [16].

Abbreviations

CAP	Cold Atmospheric Plasma
RNS	Reactive Nitrogen Species
ROS	Reactive Oxygen Species
RONS	Reactive Oxygen and Nitrogen Species

DOI: http://dx.doi.org/10.5772/intechopen.1002628

NTP	Nonthermal atmospheric pressure plasma
VUVAS	Vacuum ultraviolet absorption spectroscopy
OES	Optical emission spectroscopy
TALIF	Two-photon absorption laser-induced fluorescence
ESR	Electron spin resonance
PCR	Polymerase chain reaction
PRP	Platelet-rich Plasma
CO	Carbon monoxide
PECVD	Plasma-Enhanced Chemical Vapor Deposition
VOC	Volatile organic compound
DBD	Dielectric barrier discharge

Author details

Aamir Shahzad*, Fazeelat Hanif, Alina Manzoor and Qurat-Ul Ain Asif
Department of Physics, Government College University Faisalabad (GCUF), Faisalabad, Pakistan

*Address all correspondence to: aamirshahzad_8@hotmail.com; aamir.awan@gcuf.edu.pk

IntechOpen

© 2023 The Author(s). Licensee IntechOpen. This chapter is distributed under the terms of the Creative Commons Attribution License (http://creativecommons.org/licenses/by/3.0), which permits unrestricted use, distribution, and reproduction in any medium, provided the original work is properly cited. (cc) BY

References

[1] Shahzad A, Ahmed Z, Kashif M, Sohail A, Manzoor A, Hanif F, et al. Large Scale Simulations for Dust Acoustic Waves in Weakly Coupled Dusty Plasmas. Croatia/London, UK: INTECH Publisher; 2022

[2] Abduvokhidov D, Yusupov M, Shahzad A, Attri P, Shiratani M, Oliveira MC, et al. Unraveling the transport properties of RONS across nitro-oxidized membranes. Biomolecules. 2023;**13**(7):1043

[3] Tanaka H, Ishikawa K, Mizuno M, Toyokuni S, Kajiyama H, Kikkawa F, et al. State of the art in medical applications using non-thermal atmospheric pressure plasma. Reviews of Modern Plasma Physics. 2017;**1**:1-89

[4] Shahzad A, editor. Plasma Science and Technology. London, UK: IntechOpen; 2022. DOI: 10.5772/intechopen.95256

[5] Popescu MN, Iliescu MG, Beiu C, et al. Autologous platelet-rich plasma efficacy in the field of regenerative medicine: Product and quality control. Hindawi BioMed Research International. 2021;**2021**:4672959, 6 pages

[6] Cognasse F, Hamzeh-Cognasse H, Mismetti P, Thomas T, Eglin D, Marotte H. The nonhaemostatic response of platelets to stress: An actor of the inflammatory environment on regenerative medicine? Frontiers in Immunology. 2021;**12**:741988

[7] Carbone E, Douat C. Carbon monoxide in plasma medicine and agriculture: Just a foe or a potential friend? Plasma Medicine. 2018;**8**(1):93-120

[8] Pankaj SK, Bueno-Ferrer C, Misra NN, Milosavljević V, O'donnell C, et al. Applications of cold plasma technology in food packaging. Trends in Food Science & Technology. 2014;**35**(1):5-17

[9] Shahzad A, He M, editors. Emerging Developments and Applications of Low Temperature Plasma. Hershey, PA, US: IGI Global; 2021

[10] Randeniya LK, de Groot GJ. Non-thermal plasma treatment of agricultural seeds for stimulation of germination, removal of surface contamination and other benefits: A review. Plasma Processes and Polymers. 2015;**12**(7):608-623

[11] Kinay P, Mansour MF, Gabler FM, Margosan DA, Smilanick JL. Characterization of fungicide-resistant isolates of Penicillium digitatum collected in California. Crop Protection. 2007;**26**:647

[12] Hamaguchi S, Agarwal S, Zajickova L, Wertheimer MR. Plasmas for microfabrication. Plasma Processes and Polymers. 2019;**16**(9):1990001

[13] Graves DB. Plasma processing. IEEE Transactions on Plasma Science. 1994;**22**(1):31-42

[14] Hackam R, Akiyama H. Air pollution control by electrical discharges. IEEE Transactions on Dielectrics and Electrical Insulation. 2000;**7**:654

[15] Kim H-H. Nonthermal plasma processing for air-pollution control: A historical review, current issues, and future prospects. Plasma Processes and Polymers. 2004;**1**:91

[16] Dave H, Ledwani L, Nema SK. Nonthermal plasma: A promising green technology to improve environmental performance of textile industries. In: The Impact and Prospects of Green Chemistry for Textile Technology. Cambridge, UK: Woodhead Publishing; 2019. pp. 199-249

Chapter 2

Structural Analysis of Strongly Coupled Dusty Plasma Using Molecular Dynamics Simulation

Aamir Shahzad, Fazeelat Hanif and Alina Manzoor

Abstract

Equilibrium molecular dynamics (EMD) simulation has been used to investigate structural behaviors (order-disorder structures) of three-dimensional (*3D*) strongly coupled dusty plasmas (SCDPs). The Yukawa (screened coulomb) potential and periodic boundary conditions (PBCs) have been used in the SCDPs algorithm. Two factors have been used to analyze the structural behavior of SCDP which are radial distribution function (RDF), and lattice correlation (LC). The results for these factors have been calculated in a canonical (*NVT*) ensemble at external electric field strength ($E^* = 0.03$) for different plasma conditions of Coulomb coupling (Γ) and Debye screening parameters (κ) at the number of particles ($N = 500$). Their results have shown that the *3D* SCDP structure moves from a disordered to an ordered state with increasing Γ, and the long-range order moves to high Γ with an increase of κ. In comparison to earlier numerical, experimental, and theoretical data, the obtained results have been found to be more acceptable.

Keywords: lattice correlation, radial distribution function, dusty plasma, molecular dynamics, electric field strength

1. Introduction

Now, complex liquids have become a matter of interest for many researchers. Different techniques such as experimental, theoretical, and simulation are used to study the behavior of complex liquids. Unambiguous models have been used to describe physical properties for a specific range of temperature and pressure. Explicit equations can also be used to calculate the thermophysical properties of liquids when there is not any literature on complex liquids present. The thermophysical properties of complex liquids are altered with the change in temperature, pressure, and material composition, but the chemical properties stay unaffected. The complex liquids' phase change is described by thermophysical properties. These are divided into two categories transport and thermodynamic properties. Thermodynamic properties explain the system equilibrium conditions which comprise of entropy, heat capacity, internal energy, pressure, density, and enthalpy while the transport properties contain thermal conductivity, viscosity, RDF, waves with its instabilities, and diffusion. These transport properties state the momentum and energy transfer to the system under concern.

These properties help in system designing and have knowledge of the physical mechanism. These properties have many applications in industry and laboratory [1].

1.1 Plasma

About 99% of matter that exists in space is plasma and it is called the fourth state of matter. Mainly, plasma occurs in electrified gas form, where atoms are dissociated into positive ions and electrons. It is a form of matter in different areas of Physics such as technical plasma, terrestrial plasma, solid-state plasma, space plasma, and astrophysics. Plasma is produced artificially in laboratories used for numerous applications likely in a display, fluorescent lights, fusion energy research, and others. The term "plasma" was first time used by Irving Langmuir, who is an American physicist and he defined plasma as "plasma is a quasineutral gas of charged and neutral particles which shows collective behavior". The neutral gas ionization contains equal numbers of positive and negative charge carriers ($n_i \approx n_e \approx n$). Where n_i is ion density, n_e is electron density, and n is number density. Such plasma is called "quasi-neutral". Collective behavior means that plasma shows a simultaneous response of many particles to an external stimulus. The Debye spheres that surround small particles in plasma are distorted, which allow an externally applied electric field to regulate the interparticle interaction [2]. The main examples of plasma are planetary objects, nuclear fireballs, neon signs, plasma balls, and welding arcs. Plasma is widely used in the field of science and technology. It has a very important part in our daily life. It is used in gas lasers, sterilizing of medical instruments, industries, gas discharge, intense power beams, lightning, water purification plant, and many more [3].

1.2 Plasma history

Irving Langmuir American scientist defined plasma for the first time in 1922. Some academics also worked on plasma physics in 1930. Hans Alfven developed hydromagnetic waves in 1940; these waves are known as Alfven waves, and they were used to explore astrophysical plasma. Beginning around the same time in the Soviet Union, Britain, and the United States in 1950, magnetic fusion energy research was initiated. The study of magnetic fusion energy was regarded as a subfield of thermonuclear power in 1958. By using a Russian Tokomak design, plasma with various plasma properties was produced at the end of 1960. Numerous sophisticated tokamaks that were created between 1970 and 1980 validated the effectiveness of tokamaks. Additionally, the tokamak nearly achieved a fusion break in 1990, and the research on DP physics had begun. DP is defined as "when charged particles absorbed in plasma, becomes four components plasma containing ions, electrons, neutral and dust particles" and dust particles change the plasma properties which is called "dusty plasma" [4].

1.3 Types of plasma

The plasma can be categorized on the bases of the Coulomb coupling factor (Γ) into two categories.

1.3.1 Ideal plasma (weakly coupled plasma)

A plasma is called ideal plasma if $\Gamma \sim 10^{-4} \ll 1$. It is a kind of plasma whose temperature is very high and density is low, and in this plasma average kinetic energy

(K.E) is considerably more than the interaction potential energy (P.E) of particles, even if we can ignore this interaction P.E due to increase of average distance between the particles. It is also called as weakly coupled (ideal) plasma (WCP) because the Debye sphere is densely occupied. Because of small interaction, this plasma does not have any organization of particles or self-organization similar to plasma crystal, and plasma behaves as a nonideal gas. Its example is ordinary plasma and hot plasma [5].

1.3.2 Nonideal plasma (strongly coupled plasma)

A plasma is called nonideal plasma if $\Gamma = 10^4 \gg 1$. It is a kind of plasma whose temperature is very low and density is high and in this plasma average K.E is considerably small than the interaction P.E of particles. It is also called as SCP and cold plasma. They can have the structure of a liquid for weak to intermediate Γ values and the structure of a crystalline lattice or solid for higher Γ values. It appears in various physical systems such as electrons drifting of helium liquid on its surface, and astrophysical systems such as giant planetary interiors, neutron stars, ions in the interior of white dwarfs, dusty plasma, and cryogenic traps. An example is laboratory plasma [5].

1.4 Complex (dusty) plasmas (CDPs)

DP is a complex many-component plasmas containing ions, electrons, neutral particles, and dust particles. Some plasmas naturally contain dust particles or they may be introduced artificially through different mechanisms e.g., sputtering, etching, etc. CDPs contain dust particles of sizes ranging from tens of nanometers to hundreds of microns. The size of the particle is $3e^{-8}$g. DP properties become more complex when charged particles are absorbed in plasma and also when used on the laboratory scale. DP is called as CDP in similarity to the condensed matter field of "complex liquids" in soft matter (polymers, colloidal suspensions, surfactants, etc.). Studies of solitons, shock waves, crystallization and melting fronts, and crystal dislocations are just a few examples of the experiments that have been inspired by this concept. CDP, which can be utilized for similar reasons but differ from colloidal suspensions in having weak damping, make it possible to investigate various processes on their inherent dynamic time scale [6]. Dust particles can be in solid (metallic), liquid, gaseous, crystal, or dielectric form. On the basis of the ordering of many radii and characteristic lengths between particles interacting (r_d, λ_D), plasma with dust particles can be called either "dusty plasma" or "dust in plasma". If the $\lambda_D > r_d$ then it is called "dusty plasma" and if $\lambda_D < r_d$ then it is called "dust in plasma". Here λ_D is dust particles Debye length and r_d is the interparticle distance [4, 7].

1.5 History of dusty plasma

In 1924, the term plasma was first time defined by Irving Langmuir. In 1980, a very exciting incident happened in the field of DP for the Saturn ring and in laboratory measure of DP dynamical structure parameter was performed. Hannes Alfven and Lyman Spitzer proved dust particles as an important component and images from the Saturn ring have shown that dust particles are rotating around the Saturn ring, having the shape of spokes. The Ulysses European spacecraft passed by Jupiter in 1992, detecting dust particles and measuring their weights and collision speeds. In 1995, NASA's Galileo spacecraft detected the source of streams of dust around Jupiter.

In 1997, Mendis discovered a bright comet by a distant ancestor. The distant ancestor is an extraordinary comic laboratory for the investigation and study of interactions between dust particles and their dynamic and physical behaviors. Other displays of DP were noctilucent clouds, origin nebula, zodiac light, etc. The research from the last analysis has shown that these dust particles are fine particles. The present condition (2000–2017) of DP is stable and is playing a very dominant role in science, industries, technology, energy sectors, and medical stores [8].

1.6 Charge on dust particle

Dust particles have almost e^4 order of charge. Their charge is mostly negative in low-temperature laboratory plasma but depending on the charging process charge may be positive. Charging of dust particles can be drained through different processes involving the background plasma (electron and ion) being bombarded on the surface of dust particles, secondary electron production, ion sputtering, photoelectronic emission by UV radiations, etc. When an electron from the background plasma strikes the surface of a dust particle, it gets an electric charge. As electrons are more mobile than ions, the surface dust particle collects electrons, attracts ions, and repels electrons until the state of immobility is attained. Other collective phenomena and wave instabilities are created due to interactions between these particles. This charge is accountable for a long lifetime of particles and confinement in plasma. Dust particle potential is reliant on some parameters such as grain composition, grain size, plasma condition, temperature, and velocity [7].

1.7 Formation of dusty plasma in laboratory

In-lab manufacturing of DP has been expanded using a variety of techniques. Various techniques, including modified Q machines, DC discharges, and RF discharges, have been applied recently. The modified Q machine, a single-ended device used to produce DP, permits the dust particles to disperse throughout a portion of the cylindrical plasma column. Dust particles smaller than a micron form DP in the stratum of dc neon glow discharge. Cold electrodes in a glass tube with a cylinder shape are used to create the discharge. The electrodes are 40 cm apart. Neon pressure varies from 0.2 to 1 Torr, while discharge current varies from 0.4 to 2.5 mA. In a symmetric cylindrical rf plasma system, DP is restrained. The grounded electrodes, the hollow outer electrode, capacitively connected to a 14 MHz rf power amplifier, and the glass window make up the system of RF discharges. Thermonuclear fireballs, dust precipitators, rocket exhaust, dust in space stations, and laboratory-generated DP are examples of man-made DP. Microelectronics fabrication, Plasma Enhanced Chemical Vapor Deposition, flat-panel displays, solar cells, and semiconductor chips are among further uses for DP. DP devices (DPDs) are used in laboratories to create DP. Ordinary flames, candle flames, intense desire, fire, and blaze created by burning gas are essentially weakly ionized plasma with dust particles. The degree of ionization in typical hydrocarbon burns is increased by the thermionic electron emission of 10 nm dust particles. DP also contains in-charged snow and volcanos [9, 10].

2. Methodology

To evaluate the interactions and movements of atoms and molecules at a given time period and to investigate the time relating role of the molecular system a computational method MD simulation is used. It is reliant on Newton's second law of motion, stated as F = ma. The $F_i = \sum_j F_{ij}$ is the force experienced on all dust particles. The force on *ith* particle is exerted by other particles. In this study, each simulation has considered 500 particles for a face-centered cubic (fcc) lattice in an *NVT* ensemble. PBCs are imposed on this lattice. Particles interact with each other through Yukawa (screened Coulomb) potential. When there is no E^* Yukawa potential which is an ideal potential for CDPs study is given as

$$\phi(|\mathbf{r}|) = \frac{Q^2}{4\pi\varepsilon_0} \frac{e^{-|\mathbf{r}|/\lambda_D}}{|\mathbf{r}|} \tag{1}$$

In the above equation, r represents interparticle distance, λ_D is Debye screening length, Q is a charge on dust grains and ε_o is the permittivity of free space [5, 11].

However, particle interaction occurs under the effect of E^* whose value is 0.03, which causes electric force which is given as $F_e = Q_d E$. This force causes the drifting of negative dust particles. Basically, there are four forces which are the force of gravity, the electric force, the neutral drag force, and the thermophoretic force acting on dust particles in the direction of E^*. In this work, we have assumed that E^* makes chains or strings of plasma causing structural order resulting in plasma crystal is obtained. The drift velocity due to dust particles flow is given as $v_d = E/mv$, where m indicates mass and v is the frequency of collision with neutrals. As E^* acts along the z-direction whose z-component is given as

$$E = (0, 0, Ez) = eEr\cos\theta \tag{2}$$

where z indicates unit vector direction along the z-axis and θ gives an idea of the direction of E^*. Now, there are two terms included in interaction potential which are Yukawa potential which is repulsive interaction, and electric potential whose interaction is attractive [12].

$$\phi(|\mathbf{r}|, \theta) = \frac{Q^2}{4\pi\varepsilon_0} \frac{e^{-|\mathbf{r}|/\lambda_D}}{|\mathbf{r}|} - eE_z r\cos\theta \tag{3}$$

System particle is expressed by plasma parameters. Coulomb coupling strength (Γ) is defined as the potential energy of interaction between particles and the average kinetic energy of the particles. The Γ defines the distribution of plasma. Γ is given in this form:

$$\Gamma = \left(\frac{Q^2}{4\pi\varepsilon_o} \frac{1}{a_{WS} k_B T} \right) \tag{4}$$

In Eq. (2), T is absolute temperature, k_B is Boltzmann constant and a_{ws} represents Wigner Seitz (ws) radius and its value is $(3/4\pi n)^{1/3}$ with n representing the number density ($n = N/V$) of dust particle. Debye screening strength is the ratio of the interparticle distance a (Wigner Seitz radius) to the Debye length λ_D [1, 5, 8]. The value of κ can be represented in this form:

$$\kappa = \frac{a}{\lambda_D} \tag{5}$$

In this velocity verlet algorithm is used to allot initial values of acceleration and velocity to the atom. It is a 3D Ewald summation method for calculation of RDF, and LC. This method is used for long-range interactions in periodic systems [13]. In this section, the EMD simulation is described for RDF, and LC of SCDP for an extensive range of plasma parameters of ($1 \leq \kappa \leq 3$) and ($100 \leq \Gamma \leq 5$) along with $E^* = 0.03$ for $N = 500$.

The lack of any permanent structure characterizes the fluid condition. Nonetheless, there are well-structural correlations that may be studied experimentally to reveal essential insights regarding molecular structure. For systems that are spatially homogeneous, simply relative separation is significant, causing a sum over atom pairs,

$$g(r) = \frac{2V}{N_m^2} < \sum_{i<j} \delta(r - r_{ij}) > \tag{6}$$

and it is possible to average the function over angles without information loss. The $g(r)$ – RDF is defined as a function that defines the spherically averaged local grouping around some certain atom. The description of $g(r)$ suggests that $\rho g(r)\, dr$ is related to the probability of locating an atom in the dr volume element at a distance r from a certain atom, and in 3D, $4\pi\rho g(r) r^2 \Delta r$ is the atoms' mean number in a shell of thickness Δr and radius r surrounding atom [14]. The RDF of the system of particles (dust grains) in equilibrium shows numerous sharp peaks, which further supports the medium's crystalline state [15].

LC gives an idea of structural behavior (order-disorder (OD) structure). In this, we used the EMD technique for the simulation of long-term order and Yukawa potential. In this, we will find density at a point r by the following formula

$$\rho(r) = \sum_{j=1}^{N} \delta(r - r_j) \tag{7}$$

Long-term crystalline order will be calculated using Fourier transform

$$\psi(\kappa) = \frac{1}{N} \sum_{i=1}^{N} \exp(-ik.r_j) \tag{8}$$

The 3D Yukawa liquids (DP) motion is simulated by means of the above equation and SCCDPs OD structures can be found, and in this equation k denotes the OD position reciprocal lattice vector (RLV) of the existing Yukawa system. Here k is given as $k = 2\pi/(1,-1,1)\lambda$, when the structure is fcc then k is RLV and has been used in our EMD numerical code, where edge length is denoted by λ and $k^* = 2a_{ws}\pi/\lambda$ is the normalized value of k. The value of LC $|\Psi(k)| \sim 1$ if there is a completely ordered state of the Yukawa system, then, there is a strongly coupled system. On the other side, when the LC $|\Psi(k)| \sim 0$, there is a disordered state of the Yukawa system, that is, a gas nonideal state. Between upper and lower bounds $0 < |\Psi(k)| < 1$ the Yukawa system rests more possibly overhead the intermediary line, then, there is a moderately ordered state of the system. It is significant that the disordered state disappears with $|\Psi(k)| = O(N^{-1/2})$ [5, 14].

3. EMD simulation results

3.1 Lattice correlation as a function of time $\Psi(t)$

3.1.1 Radial distribution function as a function of distance g(r)

Computational results using EMD simulation have shown in **Figure 1** for SCDP LC, Ψ(t) as a function of time at $\kappa = 1$, 2, and 3 for four different values of $\Gamma = 5$, 20, 50,

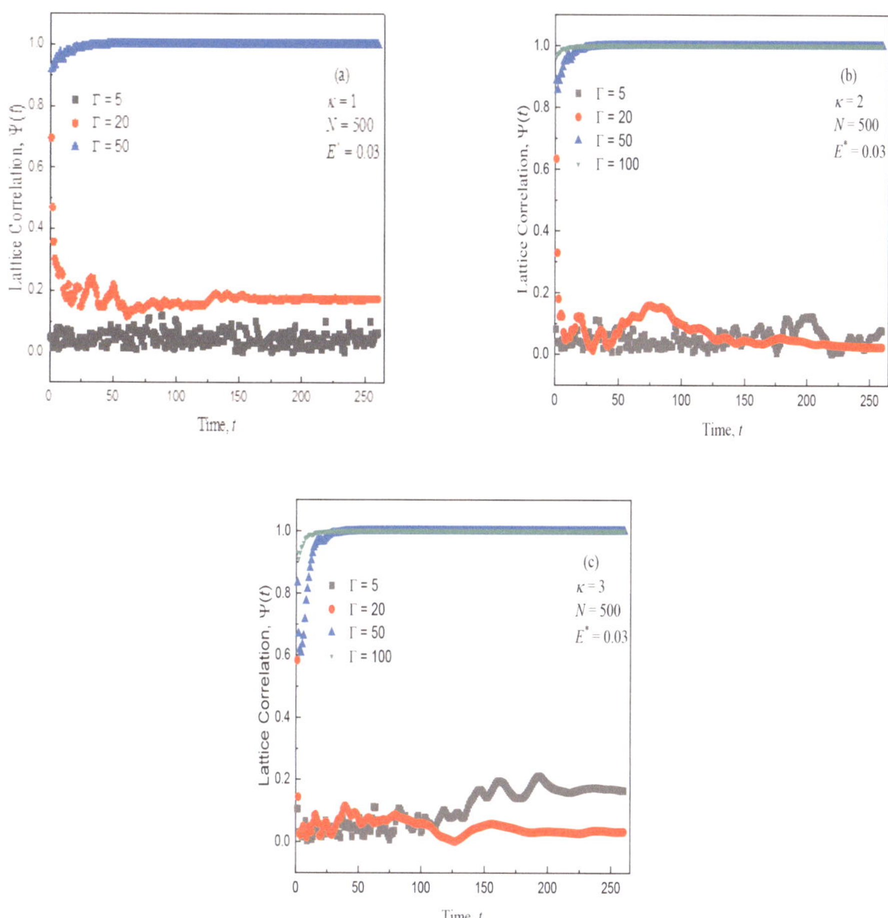

Figure 1.
The variation of LC, $\Psi(t)$ as a function of time of SCDP at system size N = 500, E^* = 0.03, and Γ = 5, 20, 50, and 100 for three different plasma factors (a) κ = 1, (b) κ = 2, and (c) κ = 3.

and 100 and system size is N = 500 and external electric field E^* = 0.03 which is normalized. Computational results using EMD simulation have shown in **Figure 2** for SCDP RDF, $g(r)$ as a function of distance at κ = 1, 2, and 3 for four different values of Γ = 5, 20, 50, and 100 and system size is N = 500 and E^* = 0.03 which is normalized and windows within graph have shown elaborated peaks. It has been obtained from **Figure 1** that long-range order moves to high Γ with increasing κ and for moderate and larger values of Γ, a moderate to high ordered structure is obtained. The collective effect is one of the reasons for the larger values of Γ (e.g., $\Gamma >> 100$), and the motion of particles is weak sufficiently that particles incline to stay caged or trapped near-equilibrium points which are centered among near neighbors and a crystalline-like lattice is formed by self-organizing of particles [5, 16, 17]. It has been obtained from **Figure 2** that κ value decreases no of peaks for moderate values of Γ which means structure shifts to disorderness. The high peak has been obtained at a small distance and the number of peaks decreases with increasing distance. It has been obtained that for moderate to higher coulomb coupling values, a moderate to high degree of order occurs as a greater no of sharp and fine peaks are formed. Because of repulsion among the particles, there is "correlation hole" for function $g(r)$ at short separations, which

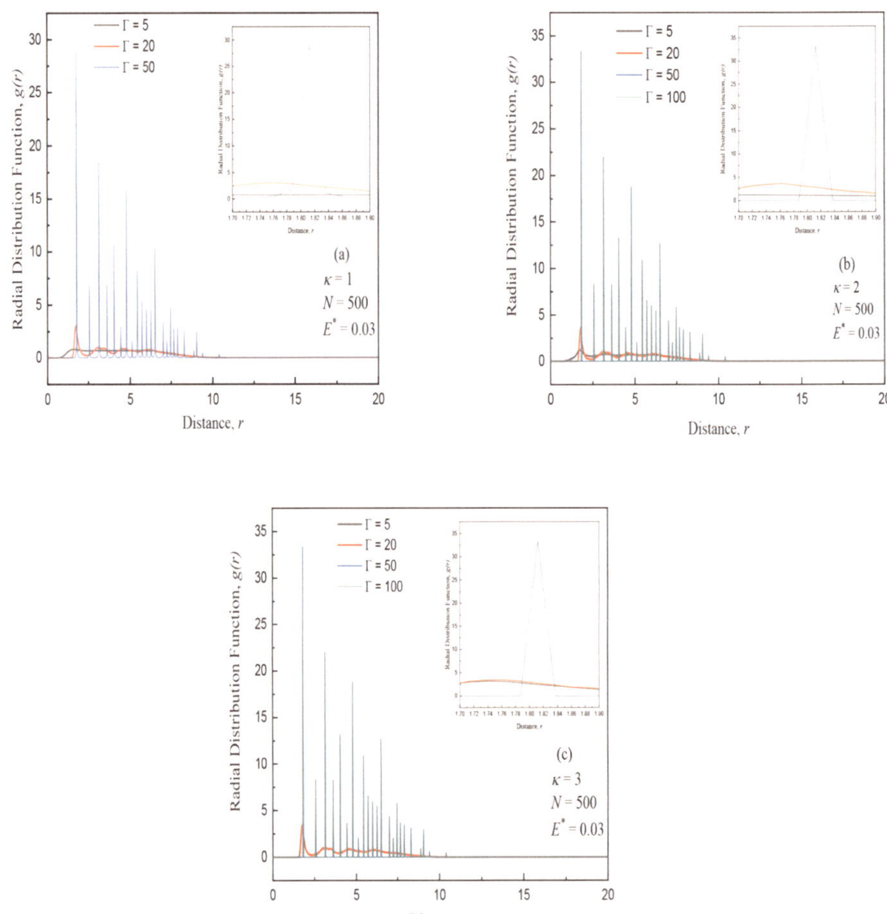

Figure 2.
The variation of RDF, $g(r)$ as a function of distance of SCDP at system size $N = 500$, $E^* = 0.03$, and $\Gamma = 5, 20, 50,$ and 100 for three different plasma factors (a) $\kappa = 1$, (b) $\kappa = 2$, and (c) $\kappa = 3$.

prohibits it from getting to close due to the absence of an adequate amount of K.E. At a distance equivalent to the most likely first neighbor distance, the first coordinating shell is marked by a peak which is exhibited by $g(r)$. This is a high point. After then, there's a depleted region and a series of other coordinated shells, which emerge in $g(r)$ in the form of smaller amplitude peaks due to correlations decay above longer distances. Fluids have a finite spatial organization across a finite distance the fluid "keeps" the lattice structure within this distance. Because of this, the $g(r)$ peak positions of $g(r)$ vary only to some extent with an increase of Γ. Because of the system's significant structure for large Γ, the particles oscillate in localized potential wells of the potential surface formed by other particles, which is known as quasi-localization. The change in the potential surface owing to particle diffusion is substantially slower as compared to the time scale of these oscillations for ($\Gamma > > 1$). The quasi-localized charge approximation is based on this behavior, with the dynamical features extracted from the interaction potential and the pair correlation function [18–20]. Screening effects have an important role in electron-electron interaction potential and atomic systems show more stability when the particular screening effects are considered [21].

DOI: http://dx.doi.org/10.5772/intechopen.1002502

4. Conclusions

Two factors LC and RDF have been used for the structural analysis of *3D* SCDPs, for a wide range of plasma parameters (Γ, κ) at E^{*} = 0.03 and N = 500 using EMD simulation. It has been obtained that the SCDP structure moves from disordered or moderate to completely ordered conditions with increasing Γ, and the long-range order moves to high Γ with an increase of κ. The presented EMD method has an outstanding performance; its consistency and accuracy are comparable to those of previous findings.

Abbreviations

DP	Dusty plasma
CDP	Complex dusty plasma
EMD	Equilibrium molecular dynamics
Γ	Coulomb coupling
κ	Debye screening strength
MD	Molecular dynamics
PBCs	Periodic boundary conditions
SCDPs	Strongly coupled dusty plasmas
WCP	Weakly coupled plasma
E^{*}	External electric field
SCP	Strongly coupled plasma
LC	Lattice correlation
RDF	Radial distribution function
PBC	Periodic boundary conditions
3D	Three-dimensional
NVT	Canonical ensemble
N	Number of particles
K.E.	Kinetic energy
P.E.	Potential energy
fcc	Face-centered cubic
OD	Order-disorder
RLV	Reciprocal lattice vector

Author details

Aamir Shahzad*, Fazeelat Hanif and Alina Manzoor
Department of Physics, Government College University Faisalabad (GCUF), Faisalabad, Pakistan

*Address all correspondence to: aamirshahzad_8@hotmail.com; aamir.awan@gcuf.edu.pk

IntechOpen

© 2023 The Author(s). Licensee IntechOpen. This chapter is distributed under the terms of the Creative Commons Attribution License (http://creativecommons.org/licenses/by/3.0), which permits unrestricted use, distribution, and reproduction in any medium, provided the original work is properly cited.

References

[1] Shahzad A, He MG. Thermal conductivity calculation of complex (dusty) plasmas. Physics of Plasmas. 2012;**19**(8):083707

[2] Ivlev AV, Brandt PC, Morfill GE, Rath C, Thomas HM, Joyce G, et al. Electrorheological complex plasmas. IEEE Transactions on Plasma Science. 2010;**38**(4):733-740

[3] Chen FF. Introduction to Plasma Physics and Controlled Fusion. Vol. 1. New York: Plenum press; 1984. pp. 19-51

[4] Bellan PM. Fundamentals of Plasma Physics. New York, US: Cambridge University Press; 2008

[5] Shahzad A, He MG. Structural order and disorder in strongly coupled Yukawa liquids. Physics of Plasmas. 2016;**23**(9): 093708

[6] Kompaneets R, Morfill GE, Ivlev AV. Interparticle attraction in 2D complex. Physical Review Letters. 2016;**116**(12): 125001

[7] Rodriguez IJ. Some Assembly Required: Computational Simulations of Dusty Plasma [University Honors Theses]. Portland, US: Portland State University; 2018. p. 622. DOI: 10.15760/honors.622

[8] Shahzad A, Shakoori MA, Mao-Gang HE, Feng Y. Numerical approach to dynamical structure factor of dusty plasmas. In: Plasma Science and Technology-Basic Fundamentals and Modern Applications. Books on Demand. Croatia/London, UK: INTECH Publisher; 2019

[9] Shukla PK, Mamun AA. Introduction to Dusty Plasma Physics. Florida, US: CRC Press; 2015

[10] Shahzad A, Khan MQ, Shakoori MA, He M, Feng Y. Thermal conductivity of dusty plasmas through molecular dynamics simulations. In: Thermophysical Properties of Complex Materials. Vol. 13. BoD – Books on Demand. Croatia/London, UK: INTECH Publisher; 2020

[11] Reshetniak VV, Filippov AV. Properties of yukawa crystals and liquid under phase equilibrium conditions. Journal of Experimental and Theoretical Physics. 2019;**129**:459-469

[12] Shakoori MA, He M, Shahzad A. Diffusion coefficients of dusty plasmas in electric field. The European Physical Journal D. 2022;**76**(11):227

[13] Shahzad A, He MG, He K. Diffusion motion of two-dimensional weakly coupled complex (dusty) plasmas. Physica Scripta. 2013;**87**(3):035501

[14] Rapaport DC. The Art of Molecular Dynamics Simulation. New York, US: Cambridge University Press; 2004

[15] Maity S, Das A, Kumar S, Tiwari SK. Interplay of single particle and collective response in molecular dynamics simulation of dusty plasma system. Physics of Plasmas. 2018;**25**(4):043705

[16] Evans DJ. Homogeneous NEMD algorithm for thermal conductivity— Application of non-canonical linear response theory. Physics Letters A. 1982; **91**(9):457-460

[17] Khrapak SA, Klumov BA, Khrapak AG. Collective modes in two-dimensional one-component-plasma with logarithmic interaction. Physics of Plasmas. 2016;**23**(5):052115

[18] Donkó I, Hartmann P, Donkó Z. Molecular dynamics simulation of a

two-dimensional dusty plasma. American Journal of Physics. 2019; **87**(12):986-993

[19] Shahzad A, Kashif M, Munir T, Perveen A, He M, Bashir S. Calculations of uniaxial tensile strength of Al–Cu–Ni based metallic glasses using molecular dynamics simulations. Physica B: Condensed Matter. 2021;**602**:412566

[20] Sukhinin GI, Fedoseev AV, Salnikov MV, Rostom A, Vasiliev MM, Petrov OF. Plasma anisotropy around a dust particle placed in an external electric field. Physical Review E. 2017; **95**(6):063207

[21] Kittner M, Klapp SH. Screening effects on structure and diffusion in confined charged colloids. The Journal of Chemical Physics. 2007;**126**(15):154902

Chapter 3

Modeling Study of Impact Effect of Chemical Reactions on Nitrogen Oxide Conversion in N_2/O_2 Mixtures under Various O_2 Concentrations

Ines Sarah Medjahdi, Abdel Karim Ferouani, Mohammed Sahlaoui and Mostefa Lemerini

Abstract

The main objective of this study is to understand the influence of various chemical reactions that participate on NO creation or reduction in N_2/O_2 mixed gas induced by negative corona discharge under different O_2 concentrations (5%, 10%, 15%, 20% and 25%). The basic chemistry of NO evolution that is presented in this study is based on a comprehensive collection of processes that were gathered into 150 specific chemical reactions involving 25 molecular, excited, atomic, and charged entities. Without the diffusion and convective factors, the density was computed using the continuity equation over a range of electric reduction fields between 50 and 90 Td ($1Td = 10^{-21}$ V.m^2), at different points in the ranges 10^{-9}–10^{-4} s. The outcomes of our numerical simulations demonstrate the impact of various chemical processes on NO production and decrease, including: $N(^2D) + O_2 \rightarrow NO + O$ and: $NO + O + N_2 \rightarrow NO_2 + N_2$ respectively. Our research has shown that at 50 and 70 Td, nitrogen oxide generation is dominated by an O_2 concentration of 5%, whereas at 90 Td, it is dominated by an O_2 concentration of 10%. These outcomes are true for both reactions.

Keywords: plasma chemistry, reaction rate, gaseous mixture, chemical kinetic, gas discharge

1. Introduction

When building and enhancing combustion systems, the emission of nitrogen oxides continues to be one of the main environmental problems [1–5]. There is a lot of research being done on gas discharge plasmas and how they might be used in other fields (physics, chemistry, biology...) [6–10]. They can be used for reforming the poisonous pollutants, such as NO_X, SO_X, CO_X, etc. These studies are based on the numerical equations for the reduction of NO_X gases in reactors. Discharges in N_2/O_2

IntechOpen

mixtures can differ considerably from discharges in pure N_2 or O_2 in a number of characteristics (principally, in chemical and ionic composition). Theoretical modeling of kinetic processes in N_2/O_2 discharges has been developed in several works [11–15].

For a better understanding of the processes taking place in the atmosphere and in a range of modern plasma technologies, the non-equilibrium kinetics of low pressure plasmas in N_2/O_2 mixtures is a crucial area of research. Understanding the operation of plasma reactors used for chemical synthesis and surface treatments of diverse materials, in particular, requires knowledge of the volume and surface kinetics of active species like N, O, $N(^2D)$, or NO [16–21]. On the subject of discharges in oxygen or air, various authors presented theoretical and experimental works [22–25]. Others were interested to study electron swarm parameters in N_2/O_2 mixtures which are of great interest in atmospheric physics, pollution control or for application as electrical insulation media [26–29]. These study at investigating the physical behavior of N_2 and O_2 mixtures in order to predict and optimize their electric strength.

I Stefanović et *al.* [30], have studied kinetics of ozone and nitric oxides in dielectric barrier discharges in O_2/NO_x and $N_2/O_2/NO_x$ mixtures. They have measured various concentrations of NO, NO_2, NO_3, N_2O_5, and O_3 by classical absorption spectroscopy in dielectric barrier discharges in flowing O_2/NO_x and $N_2/O_2/NO_x$ mixtures. They have reported on a comparison of the measured concentrations with a zero-dimensional numerical kinetic model. They have shown that their model accurately describes the kinetics of ozone and the nitric oxides in the DBD in oxygen and dry air with small admixtures of NOx. Xing Fan et *al.* [31], have been developed a coaxial dielectric barrier discharge (DBD) reactor for plasma and plasma-catalytic conversion of dilute N_2O in N_2 and N_2/O_2 mixtures at both room and high temperature (300°C). They show that N_2O conversion increases with the increase of discharge power and decreases with the increase of O_2 content. They also show that increasing the inlet N_2O concentration from 100 to 400 ppm decreases the conversion of N_2O under an N_2 atmosphere but increases that under an N_2/O_2 atmosphere. They have concluded that concentrating N_2O in the N_2/O_2 mixture could alleviate the negative influence of O_2 by increasing the involvement of plasma reactive species (e.g., $N_2(A^3\Sigma^+)$ and $O(^1D)$) in N_2O conversion. Haefliger et *al.* [32], have studied experimentally derived rate coefficients for electron ionization, attachment and detachment as well as ion conversion in pure O_2 and N_2/O_2 mixtures. Four ion species and seven rate coefficients make up the model they are utilizing. They compared their findings to a number of sources, and most rate coefficients showed good agreement. They demonstrate that the waveforms' informative content is sufficient to differentiate between the two detaching ions, O and O^-, and their method may be used to analyze other gases or mixes that undergo electron detachment and ion conversion. For estimating the critical electric field strength of the mixtures, they propose a criterion based on the eigenvalues of the matrix describing the interaction of electrons and ions. Yamamoto et *al.* [33] developed a method for the breakdown of NOx pollutants that involved a chemical reactor coupled with plasma discharge. Since the plasma nitrogen monoxide was transformed into nitrogen dioxide, the reduction mechanisms' action resulted in the transformation of nitrogen dioxide into nitrogen. They looked at three different plasma reactor models, and they noticed that nearly all of the nitrogen dioxide was breaking down. Mei-Xiang et *al.* [34] investigated metal catalyst and plasma discharge combination for simultaneous removal of NOx and dust from diesel exhaust. During the plasma creation, the temperature for combustion falls and the efficiency of NOx to NO conversion is reduced. Wang et al. [35] employed the DBD method to reform carbon dioxide inside the plasma reactor and create nonthermal plasma. Separate studies on the reformation of carbon dioxide by catalytic and nonthermal plasma were conducted. They discovered that employing a DBD reactor significantly reduces CO_2 concentration over time compared to other techniques. Abedi-Varaki et *al*. [36] studied the temporal fluctuations of the number density and reaction rate of the nitrogen monoxide molecules and other generated species inside the DBD plasma reactor using a zero-dimensional model.

In this way, the aim of the present study is to simulate various values of O_2 concentrations (5%, 10%,15%, 20% and 25%), the temporal evolution of 25 chemical species (electrons e, molecules N_2, O_2, O_3, atoms N, O, nitric oxides NO, N_2O, NO_2, NO_3, N_2O_5, positive ions (O^+, O_2^+, O_4^+, N^+, N_2^+, NO^+), negative ions (O^-, O_2^-, O_3^-, O_4^-, NO_2^-, NO_3^-) and metastables species $N(^2D)$, $O(^1D)$. These different species react following 150 selected chemical reactions. The time varied from 10^{-9} to 10^{-4} s. In this numerical simulation we suppose various effects induced by the passage of a corona discharge in a mixed gas. For the sake of simplification, we assume that the gas has no convective movement gradients and the temperature remains constant.

2. Basic formulas

The basic formulas used in the present work consists of a system of equations that takes into account the variation of the density and the chemical kinetics of the environment. We developed a zero order numerical code to resolve the transport equations for neutral and charged particles. The algorithm is based on the time integration of the system of equations under consideration the variation of the density and the chemical kinetics of the environment. The system of the chemical kinetics equations can be described by a system of ordinary differential equations (i.e., the algorithm is defined by time integration) of the following form:

$$\frac{dN_i}{dt} = \sum_{j=1}^{j_{max}} S_{ij} \quad \text{where } j = 1, \dots, j_{max} \tag{1}$$

where:

$$S_{ij} = \left(G_{ij} - L_{ij}\right) \tag{2}$$

N_i show the vector of species densities, i = 1 up to 25 thought about in the plasma and S_{ij} the source term vector related to the contributions from various processes and dependent on the reaction coefficients. G_{ij} and L_{ij} stand for the gain and loss of species i as a result of chemical reactions from j = 1 to j_{max} = 150, respectively. The solution of such a system requires the knowledge of the initial concentrations. The total density N of the gas is given by the ideal gas law:

$$P = Nk_bT \tag{3}$$

Where P displays the pressure, k_b Boltzmann constant and T the absolute temperature. The source term S_{ij} of the density conservation factored in the gas's reactivity Eq. (1).

$$G_{ij} = \sum_{\mu} K_{\mu}(T)\left(n_i n_j\right)_{\mu} \quad (4)$$

$$L_{ij} = \sum_{\nu} K_{\nu}(T)\left(n_i n_j\right) \quad (5)$$

$K\mu(T)$ and $K\nu$ (T) are the reaction's coefficients μ or ν and $(n_i n_j)$ is the product of the species i and j interaction densities with the reaction μ or ν. This Arrhenius formula is satisfied by these coefficients.

$$K_{\mu}(T) = A.exp\left(-\theta_{\mu}/T\right) \quad (6)$$

$$K_{\nu}(T) = B.exp\left(-\theta_{\nu}/T\right) \quad (7)$$

Where A and B are the constants factor and $\theta\mu$ and $\theta\nu$ are the reactions activation energy and T is the species absolute temperature.

3. Results and discussion

The chemical kinetics involves 25 different chemical species: electrons (e), molecules N_2, O_2, O_3, atoms N, O, nitric oxides NO, N_2O, NO_2, NO_3, N_2O_5, positive ions (O^+, O_2^+, O_4^+, N^+, N_2^+, NO^+), negative ions (O^-, O_2^-, O_3^-, O_4^-, NO_2^-, NO_3^-), metastable species $N(^2D)$, $O(^1D)$. These different species react following 150 selected chemical reactions, the main ones are given in **Table 1**.

In this section we will analyze the effects of important chemical reactions that participated on creation or reduction of NO specie. In our work, we have found that the following reactions are the most influential on creating NO:

$$N(^2D) + O_2 \rightarrow NO + O \quad (8)$$

$$N + O_2 \rightarrow NO + O \quad (9)$$

$$O_3^- + N \rightarrow NO + O_2 + e \quad (10)$$

$$O_2^+ + N_2 \rightarrow NO + NO^+ \quad (11)$$

$$O_2^- + NO^+ \rightarrow NO + O_2 \quad (12)$$

and the others following reactions are the most influential on reducing NO:

$$NO + O + N_2 \rightarrow NO_2 + N_2 \quad (13)$$

$$NO + N \rightarrow N_2 + O \quad (14)$$

$$NO + O_3^- \rightarrow NO_3^- + O \quad (15)$$

$$NO + O(^1D) \rightarrow O_2 + N \quad (16)$$

$$NO + O_3 \rightarrow NO_2 + O_2 \quad (17)$$

So, we proposed to represent only the most important reactions namely (R1) and (R10). For that, we analyze for each electric reduced field (50, 70 and 90 Td), the evolution of density and reaction rate between 10^{-9} and 10^{-4} s, under different O_2 concentrations: 5%, 10%,15%, 20% and 25%.

	Chemical reactions	Rate coefficients
R1	$N(^2D) + O_2 \rightarrow NO + O$	$k_1 = 0.97 \times 10^{-11}$
R2	$N + O_2 \rightarrow NO + O$	$k_2 = 0.44 \times 10^{-11}$
R3	$O_3^- + N \rightarrow NO + O_2 + e$	$k_3 = 4.31 \times 10^{-7}$
R4	$O_2^+ + N_2 \rightarrow NO + NO^+$	$k_4 = 1.0 \times 10^{-17}$
R5	$O_2^- + NO^+ \rightarrow NO + O_2$	$k_5 = 4.0 \times 10^{-7}$
R6	$N_2O_5 + NO_2^- \rightarrow NO + NO_3^- + NO_3$	$k_6 = 7.0 \times 10^{-10}$
R7	$O(^1D) + NO_2 \rightarrow NO + O_2$	$k_7 = 0.14 \times 10^{-9}$
R8	$N^+ + NO_2 \rightarrow NO + NO^+$	$k_8 = 5.00 \times 10^{-10}$
R9	$O_3 + N_2^+ \rightarrow N_2 + O + O_2^+$	$k_9 = 1.00 \times 10^{-10}$
R10	$NO + O + N_2 \rightarrow NO_2 + N_2$	$k_{10} = 2.5 \times 10^{-10}$
R11	$NO + N \rightarrow N_2 + O$	$k_{11} = 0.32 \times 10^{-10}$
R12	$NO + O_3^- \rightarrow NO_3^- + O$	$k_{12} = 1.0 \times 10^{-10}$
R13	$NO + O(^1D) \rightarrow O_2 + N$	$k_{13} = 1.5 \times 10^{-10}$
R14	$NO + O_3 \rightarrow NO_2 + O_2$	$k_{14} = 0.18 \times 10^{-11}$
R15	$NO + O \rightarrow N + 2O$	$k_{15} = 0.18 \times 10^{-11}$
R16	$NO + O_3^- \rightarrow NO_2^- + O_2$	$k_{16} = 2.6 \times 10^{-12}$
R17	$NO + O_4^- \rightarrow NO_3^- + O_2$	$k_{17} = 2.5 \times 10^{-10}$
R18	$NO + O_2^+ \rightarrow NO^+ + O_2$	$k_{18} = 3.0 \times 10^{-10}$
R19	$NO + O^+ \rightarrow NO^+ + O$	$k_{19} = 1.0 \times 10^{-12}$
R20	$NO + O^- \rightarrow NO_2 + e$	$k_{20} = 2.6 \times 10^{-10}$
R21	$O_4^+ + O^- \rightarrow O_2 + O_3$	$k_{21} = 4.00 \times 10^{-7}$
R22	$N_2O + O^+ \rightarrow N_2 + O_2^+$	$k_{22} = 2.00 \times 10^{-11}$
R23	$O_4^- + O \rightarrow O_3^- + O_2$	$k_{23} = 4.00 \times 10^{-10}$
R24	$O_4^+ + NO \rightarrow NO^+ + 2O_2$	$k_{24} = 1.00 \times 10^{-10}$
R25	$e + O_2 \rightarrow O + O(^1D) + e$	$k_{25} = 3.2 \times 10^{-11}$
R26	$e + N_2 \rightarrow N + N + e$	$k_{26} = 2.0 \times 10^{-11}$

Table 1.
The primary plasma processes that produce the primary species involved in NO reduction and production, as well as their rate coefficients (rate coefficients are in units of $cm^3 \cdot molecule^{-1} \cdot s^{-1}$ for two body reactions, and cm^6. Molecule$^{-1} \cdot s^{-1}$ for three body reactions). They are taken from the literature [37–40].

Figures 1–3, show the temporal evolution of density of $N(^2D)$ specie which participated on the creation of nitrogen oxide in N_2/O_2 mixture at 50Td under different O_2 concentrations: 5%, 10%,15%, 20% and 25%. As illustrated on these curves, the evolution of NO density depends of O_2 concentrations. However, we notice that when the O_2 concentration increases NO density decreases, except for the value 90Td where we note a decrease in the density at the concentration 5%. Globally, we observe on these three figures, a first step which goes up to 4×10^{-7} s where the density increases, and a second stage which goes up to 10^{-4} s where the density decreases.

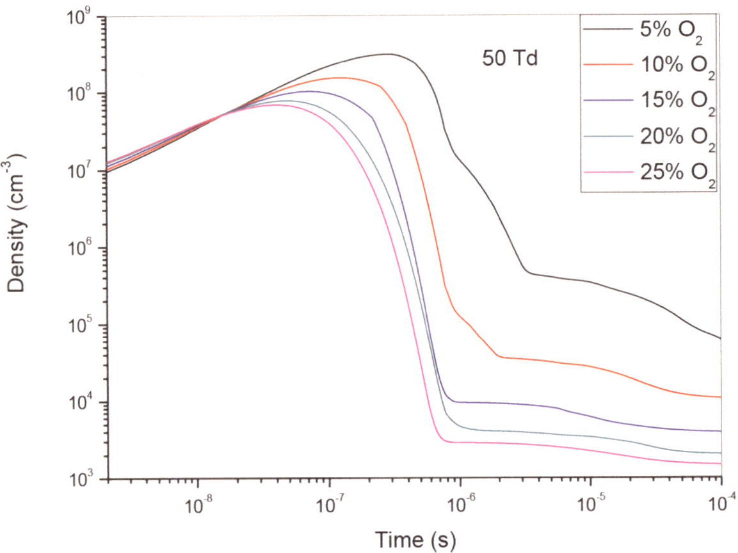

Figure 1.
Temporal evolution of N(2D) density that participate in the creation of NO specie in the mixture N_2/O_2 at 50Td under different O_2 percentages: 5%, 10%,15%, 20% and 25%.

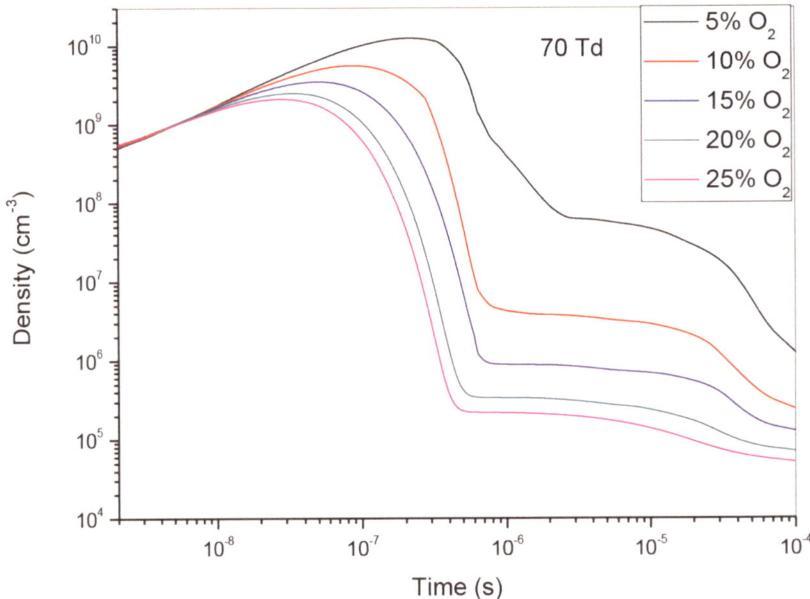

Figure 2.
Temporal evolution of $N(^2D)$ density that participate in the creation of NO specie in the mixture N_2/O_2 at 70Td under different O_2 percentages: 5%, 10%,15%, 20% and 25%.

We also notice that the gap between the different curves decreases when the oxygen concentration increases.

So, in **Figures 4–6**, we have represented the temporal evolution of O density of the same O_2 concentrations as before. We note firstly an increase of density until 10^{-6} s,

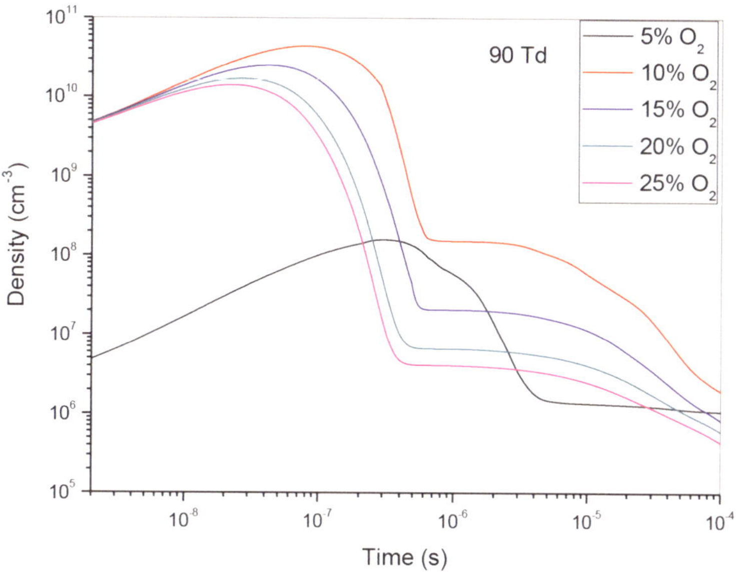

Figure 3.
Temporal evolution of N(^{2}D) density that participate in the creation of NO specie in the mixture N_2/O_2 at 90Td under different O_2 percentages: 5%, 10%,15%, 20% and 25%.

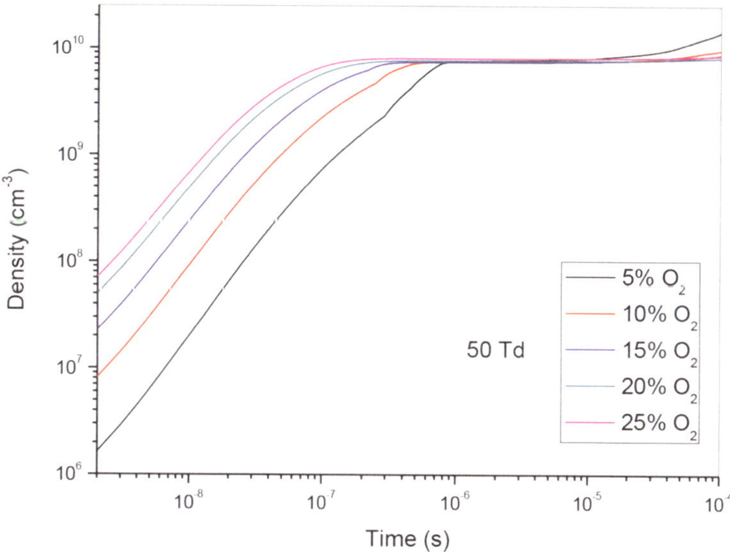

Figure 4.
Temporal evolution of O density that participate in the creation of NO specie in the mixture N_2/O_2 at 50Td under different O_2 percentages: 5%, 10%,15%, 20% and 25%.

followed by stabilization. Secondly, we observe that the rise of O_2 concentration induces the increase of atomic oxygen density.

Let us go on to analyze the reaction rate with the same values as before. We have represented in **Figures 7–9**, the temporal evolution of reaction rate of (R1): N (^{2}D) + $O_2 \rightarrow$ NO + O, that participate in the production of NO specie in N_2/O_2 mixture at 50, 70 and 90 Td, under five various O_2 concentrations: 5%, 10%, 15%, 20% and

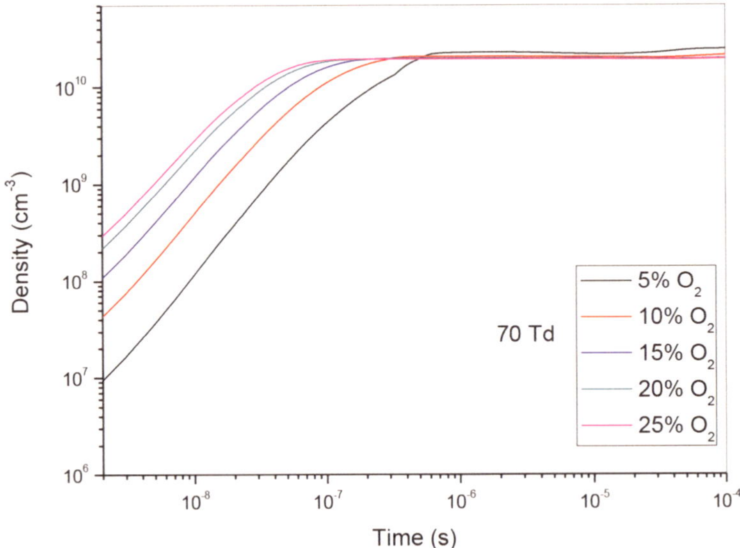

Figure 5.
Temporal evolution of O density that participate in the creation of NO specie in the mixture N_2/O_2 at 70Td under different O_2 percentages: 5%, 10%,15%, 20% and 25%.

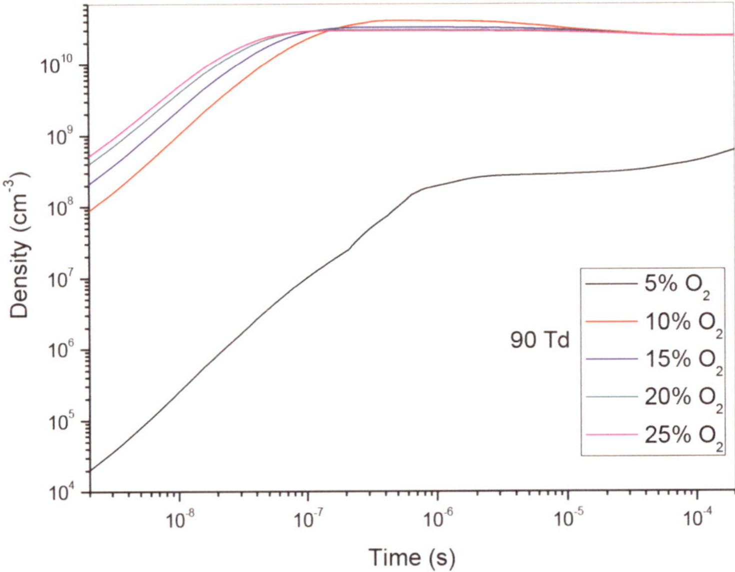

Figure 6.
Temporal evolution of O density that participate in the creation of NO specie in the mixture N_2/O_2 at 90Td under different O_2 percentages: 5%, 10%,15%, 20% and 25%.

25%. Let us start with 50 Td. We can say through curves represented in **Figure** 7, that the effect of the reaction passes by three stages:

- Between 10^{-9} and 10^{-7} s, the influence of the reaction increases with the increase of the O_2 concentrations and his domination is obtained for 25%.

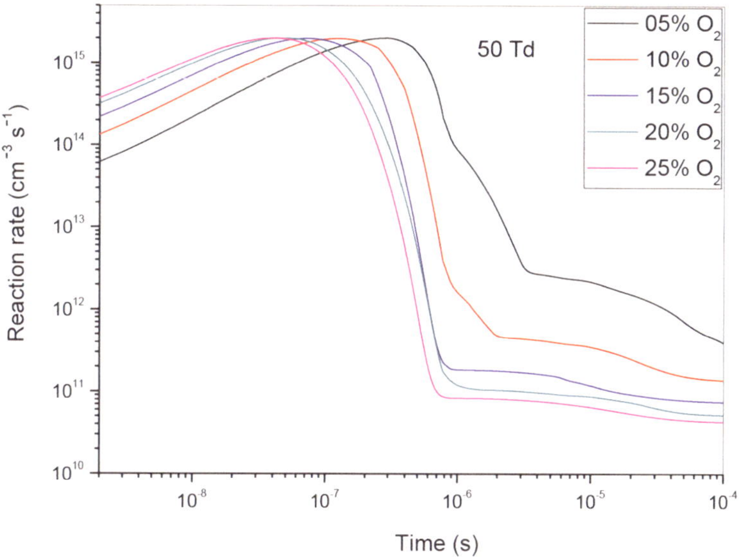

Figure 7.
Temporal evolution of reaction rate of (R1): $N(^2D) + O_2 \rightarrow NO + O$, which contributed in the creation of NO specie in N_2/O_2 mixture at 50Td, under various O_2 concentrations: 5%, 10%,15%, 20% and 25%.

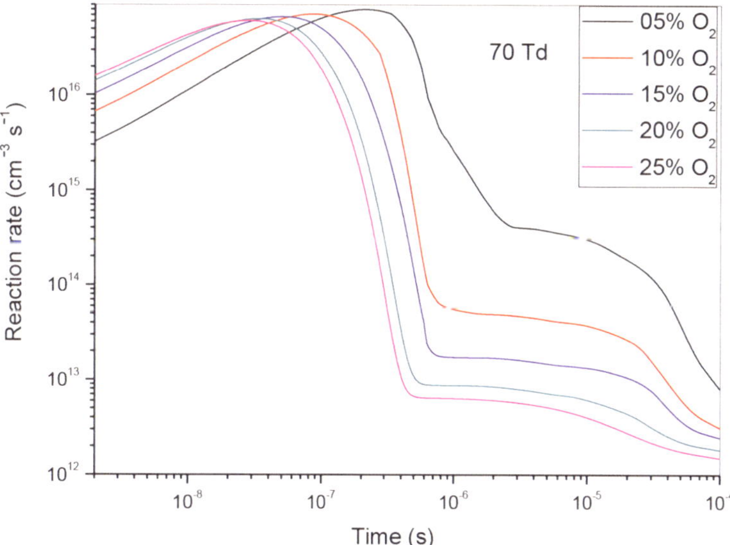

Figure 8.
Temporal evolution of reaction rate of (R1): $N(^2D) + O_2 \rightarrow NO + O$, which contributed in the creation of NO specie in N_2/O_2 mixture at 70Td, under various O_2 concentrations: 5%, 10%,15%, 20% and 25%.

- From 10^{-7} to 10^{-6} s, the reaction impact decreases quickly while the O_2 concentrations increases and his domination is obtained for 5%.

- From 10^{-6} to 10^{-4} s, the reaction's impact lessens but only somewhat.

Let us examine what occurs when the electric reduced field is increased to 70 Td. **Figure 8** presents the findings. We see that the curves have the same form as those

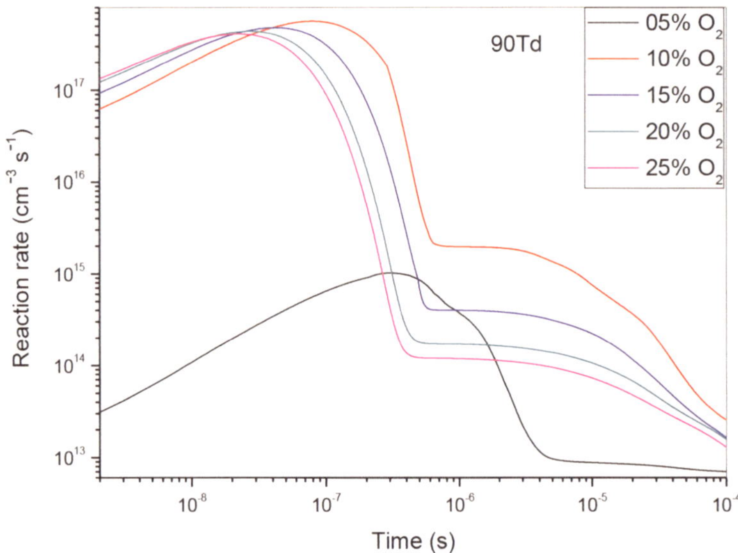

Figure 9.
Temporal evolution of reaction rate of (R1): $N(^2D) + O_2 \rightarrow NO + O$, which contributed in the creation of NO specie in N_2/O_2 mixture at 90Td, under various O_2 concentrations: 5%, 10%,15%, 20% and 25%.

previously obtained for 50 Td, but the average reaction time is shorter because of the energy injected into the gas, which is more significant.

To wrap up our research, we have demonstrated in **Figure 9**, the temporal evolution of reaction rate of (R1) which contribute to NO production at 90Td. As before, we observe the same effect for the reaction (R1), excepted the value 5%. Indeed, at this value the reaction rate does not have a great influence on the nitrogen oxide production. So, we can say that at 90 Td it's 10% of O_2 concentration that becomes dominant.

Let us now analyze the influence of reaction (R10) which is the principal reaction in the consumption of nitrogen oxide. For this reason, we have shown in **Figures 10–12** the evolution of the reaction rate for three values of the reduced electric field (50, 70 and 90 Td), and five values of O_2 concentrations (5%, 10%, 15%, 20% and 25%). We notice that NO, can react with the oxidizing radical such as O to form especially NO_2. We note firstly that the effectiveness of these concentrations is higher at the end than at the beginning. Secondly, we observe two stages of evolution:

- At 50 and 70 Td, and between 10^{-9} and 4×10^{-7} s, plus the value of the O_2 concentrations increases more reaction is effective.

- At 50 and 70 Td, and between 4×10^{-7} and 10^{-4} s, the phenomenon reverses and we observe plus the O_2 concentration increases, the effect of the reaction decreases.

- At 90 Td, the phenomena are the same as described before, with the exception of 5%, where the reaction's influence is much weaker than it is at other concentrations. In our opinion, the synthesis of the oxygen atom is insignificant at this energy level and low concentration when compared to other concentrations 10, 15, 20 and 25%.

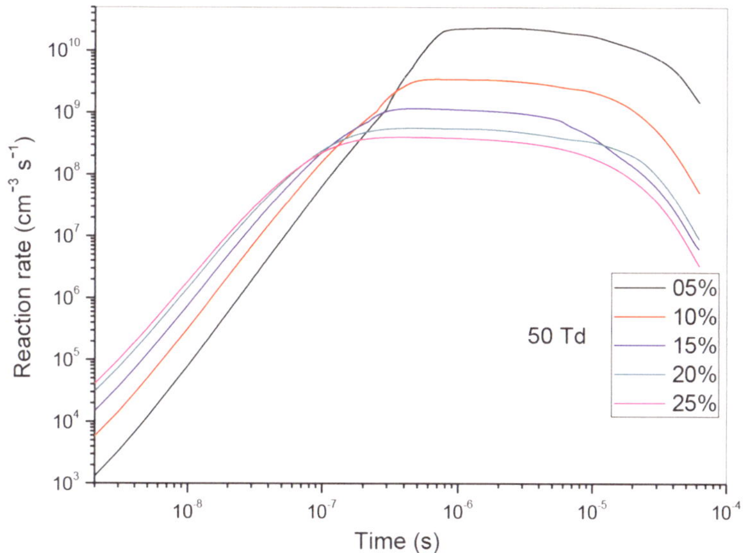

Figure 10.
Temporal evolution of reaction rate of (R10): NO + O + N_2 → NO_2 + N_2, reaction that participate in the reduction of NO specie in N_2/O_2 mixture at 50Td, under various O_2 concentrations: 5%, 10%,15%, 20% and 25%.

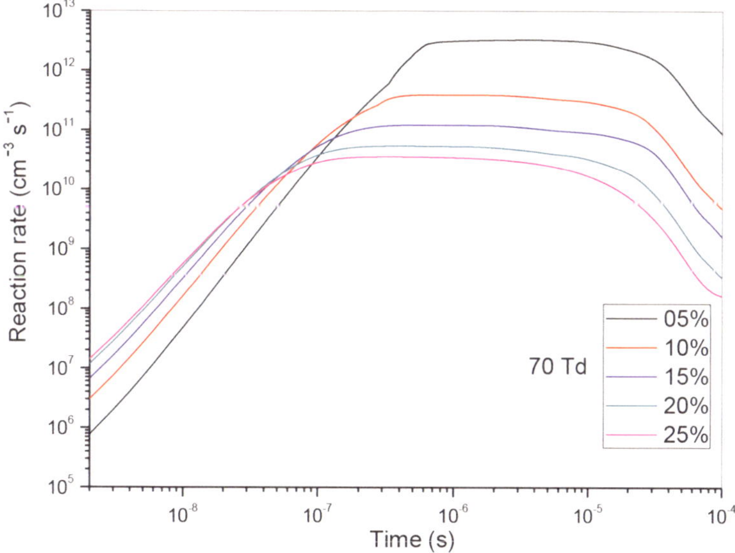

Figure 11.
Temporal evolution of reaction rate of (R10): NO + O + N_2 → NO_2 + N_2, reaction that participate in the reduction of NO specie in N_2/O_2 mixture at 70Td, under various O_2 concentrations: 5%, 10%,15%, 20% and 25%.

To highlight our results, we chose to represent them as a percentage and histogram representations. So, we have represented in the **Figures 13–18** the percentage of the two reactions that participated on the creation and reduction of nitrogen oxide to get an idea about their impact in this conversion (At left: (a) with percentage representation; At right (b) with histogram representation). First of all, let us see **Figure 13(a)**

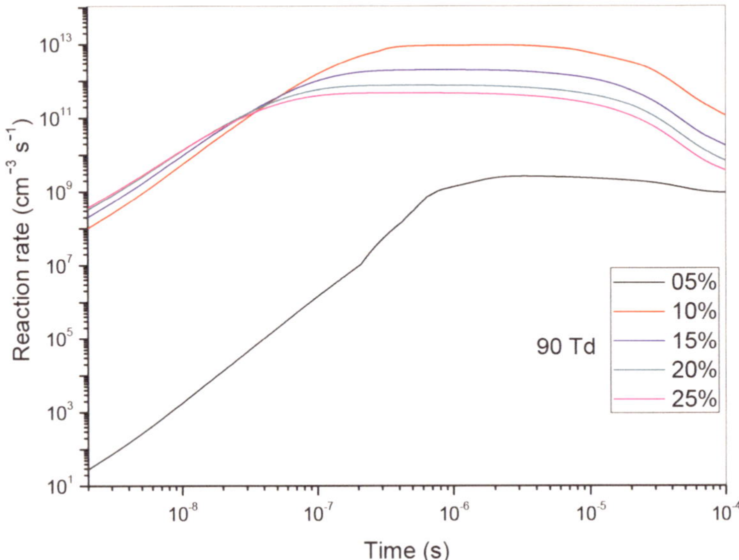

Figure 12.
Temporal evolution of reaction rate of (R10): NO + O + $N_2 \rightarrow NO_2 + N_2$, reaction that participate in the reduction of NO specie in N_2/O_2 mixture at 90Td, under various O_2 concentrations: 5%, 10%,15%, 20% and 25%.

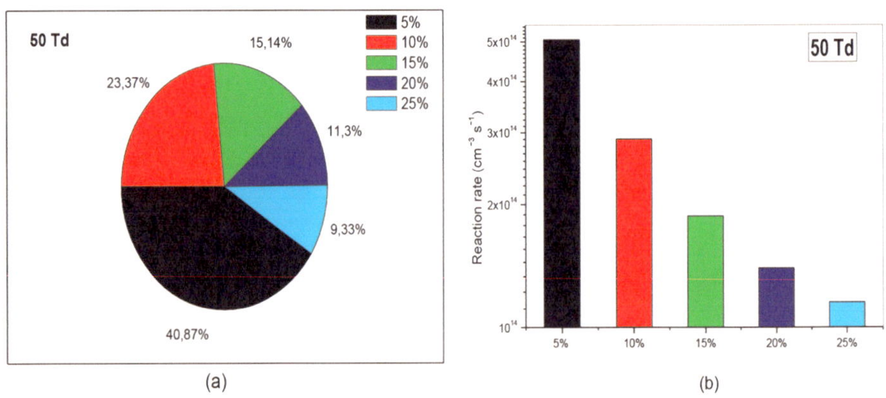

Figure 13.
Representation of reaction rate of (R1): $N(^2D) + O_2 \rightarrow NO + O$, that participate in the creation of NO specie in N_2/O_2 mixture at 50Td under various O_2 concentrations (5%, 10%, 15%, 20%, and 25%).

where we have highlighted percentages for each O_2 concentration at 50 Td. We notice that at the concentration 5%, nitrogen oxygen is crated almost four times more than when the concentration is at 25%.

Move us on to the next **Figure 14(a)**, where we have represented the percent influence for 70 Td. We observe also almost the same phenomenon as before. Indeed, we note that the creation of NO specie is still important for the concentration 5%, where we get 48.4%, against 6.96% for the concentration 25%. These results are also verified on the **Figures 13(b)** and **14(b)**, where we can say the phenomenon of NO creation dominates when the O2 concentration equals 5%.

Now let us see what happens when we increase the reduced electric field to 90 Td. The results are given in the **Figure 15(a)** and **(b)**. We clearly observe that when the

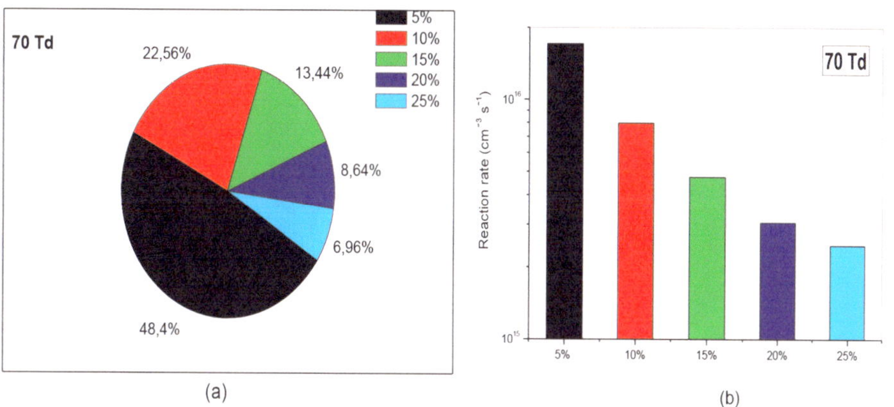

Figure 14.
Representation of reaction rate of (R1): $N(^2D) + O_2 \rightarrow NO + O$, that participate in the creation of NO specie in N_2/O_2 mixture at 70Td under various O_2 concentrations (5%, 10%, 15%, 20%, and 25%).

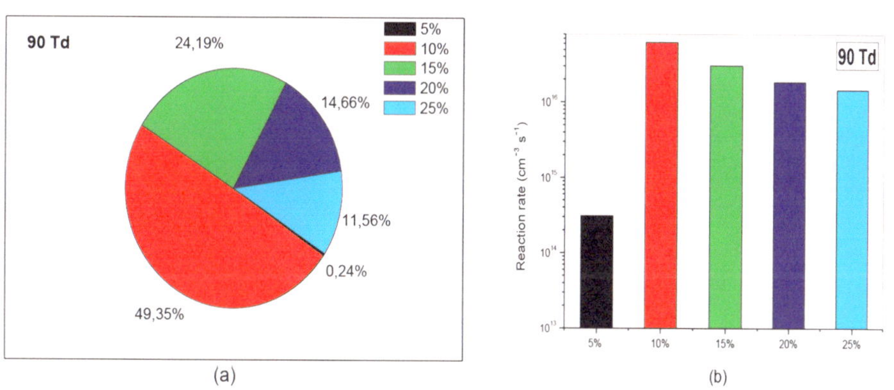

Figure 15.
Representation of reaction rate of (R1): $N(^2D) + O_2 \rightarrow NO + O$, that participate in the creation of NO specie in N_2/O_2 mixture at 90Td under various O_2 concentrations (5%, 10%, 15%, 20%, and 25%).

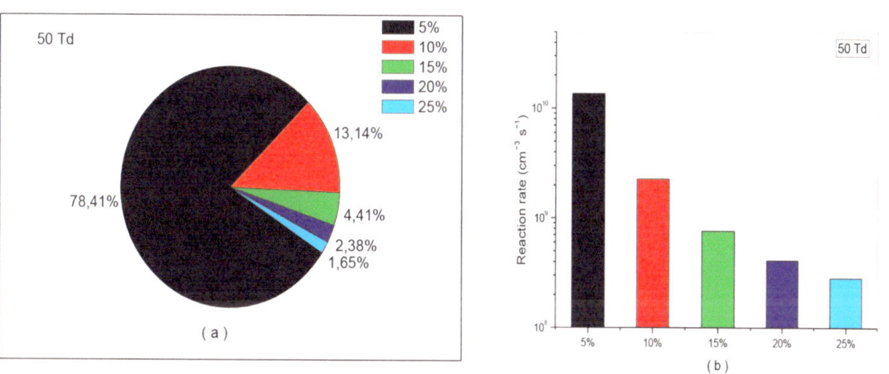

Figure 16.
Representation of reaction rate of (R10): $NO + O + N_2 \rightarrow NO_2 + N_2$, that participate in the creation of NO specie in N_2/O_2 mixture at 50Td under various O_2 concentrations (5%, 10%, 15%, 20%, and 25%).

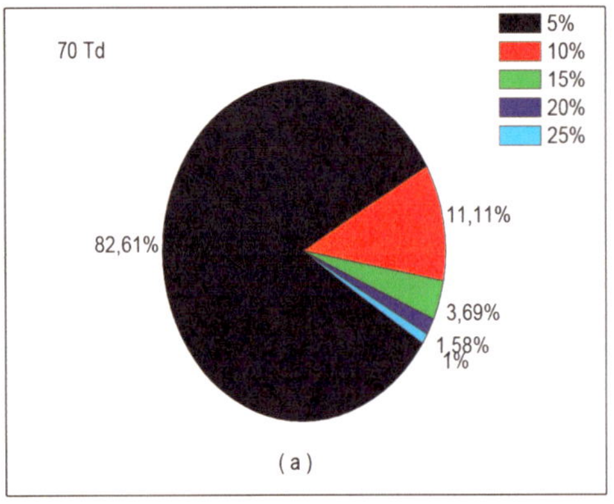

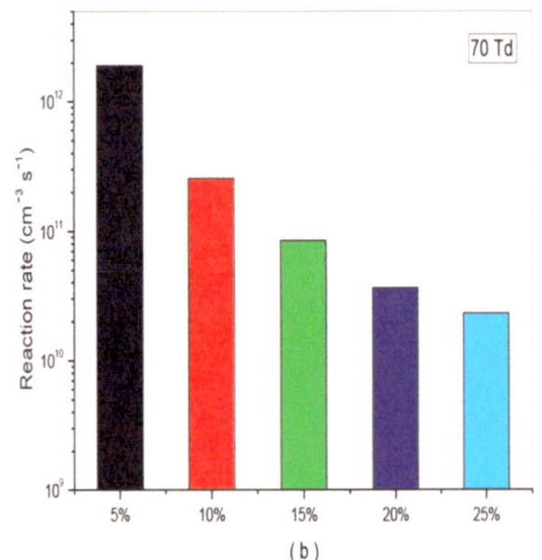

Figure 17.
Representation of reaction rate of (R10): $NO + O + N_2 \rightarrow NO_2 + N_2$, that participate in the creation of NO specie in N_2/O_2 mixture at 70Td under various O_2 concentrations (5%, 10%, 15%, 20%, and 25%).

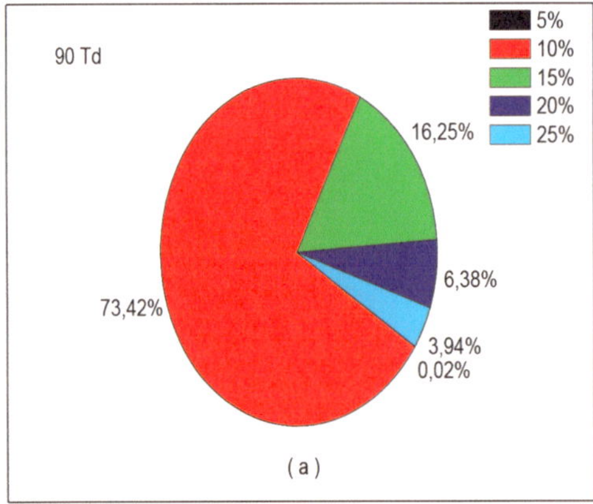

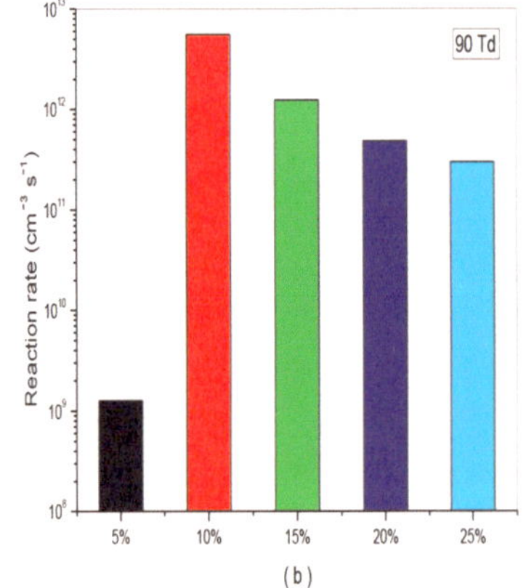

Figure 18.
Representation of reaction rate of (R10): $NO + O + N_2 \rightarrow NO_2 + N_2$, that participate in the creation of NO specie in N_2/O_2 mixture at 90Td under various O_2 concentrations (5%, 10%, 15%, 20%, and 25%).

concentration equals 5%, the creation is totally negligible (0.24%) compared to the other concentrations. Now, it's the 10% concentration that takes over and becomes dominant over all other concentrations.

Let us continue our analysis with reaction (R10) which is responsible for consumption of nitrogen oxide. The effects of this reaction have been represented in **Figures 16–18** with percentage (a) and (b) with diagram of different reactions rate. We observe also on these figures the same effect as that observed in the case of creation. The only difference lies in the intensity of the values. For example, at 50 and 70 Td we found in average 80% instead of 44%, and at 90Td we get 73.42% against 49.35%. These results are also verified on **Figures 16(b), 17(b)** and **18(b)**.

4. Conclusion

In this study, a zero-dimensional model is considered for numerical studies of NO creation and reduction in N_2/O_2 mixture under various O_2 concentrations

(5%, 10%,15%, 20% and 25%). In particular, we have analyzed in the ranges 10^{-9}–10^{-4} s the influence of chemical reactions, which actively participate in the conversion of nitrogen oxide specie at 50, 70 and 90Td. $N(^2D)$ and O are the two species on which we have concentrated.

The obtained data demonstrate that nitrogen oxide conversion varies with oxygen concentration. In fact, it has been found that the rise and fall of various species varies and are highly influenced by the values of O_2 concentrations. These findings enable us to recognize the crucial function that the reaction has in NO conversion, and they may be summed up as follows:

i. $N(^2D)$ specie produces most of the NO through the following reaction:

   $N(^2D) + O_2 \rightarrow NO + O$.

ii. The mechanism of NO reduction is due mainly by O specie through the reaction:

   $NO + O + N_2 \rightarrow NO_2 + (N_2)$

iii. We observe that the influence of the reaction (R1) was acquired for the low O_2 concentration of 5% at 50 and 70 Td for creation. In fact, we discovered an average of 44%, compared to 10% at 90 Td, where we got close to 50%.

iv. In terms of reduction, we also see that at 50 and 70 Td, the reaction's effect (R10) was attained for the low O_2 concentration of 5%, although we had typically attained 80%. The influence of this reaction was discovered for 90 Td at a concentration of 10%, which is equal to 73.42%.

Finally, the acquired data unmistakably demonstrate the significance of chemical processes and oxygen concentrations in the generation or removal of nitrogen oxide.

Acknowledgements

The work of Pr Abdel Karim Ferouani was supported by DGRSDT, Algerian Ministry of Higher Education and Research, under Project PRFU-code A11N01EP130220230001.

Author details

Ines Sarah Medjahdi[1*], Abdel Karim Ferouani[1,2*], Mohammed Sahlaoui[1,2] and Mostefa Lemerini[1]

1 Theoretical Physics Laboratory, Faculty of Sciences, Department of Physics, University of Tlemcen, Algeria

2 Higher School of Applied Sciences, ESSA-Tlemcen, Tlemcen, Algeria

*Address all correspondence to: inesmedjahdi@yahoo.com and ferouani_karim@yahoo.fr

IntechOpen

© 2023 The Author(s). Licensee IntechOpen. This chapter is distributed under the terms of the Creative Commons Attribution License (http://creativecommons.org/licenses/by/3.0), which permits unrestricted use, distribution, and reproduction in any medium, provided the original work is properly cited.

References

[1] Glarborg P, Miller JA, Ruscic B, Klippenstein SJ. Modeling nitrogen chemistry in combustion. Progress in Energy and Combustion Science. 2018; **67**:31-68. DOI: 10.1016/j.pecs.2018.01.002

[2] Zhang Y, Mathie O, Petersen EL, Bourque G, Curran HJ. Assessing the predictions of a NOx kinetic mechanism on recent hydrogen and syngas experimental data. Combustion and Flame. 2017;**182**:122-141. DOI: 10.1016/j.combustflame.2017.03.019

[3] Ferouani AK, Lemerni M, Belhour S, Askri S. Simulacion numerica de la evolucion radial y axial del campo electrico reducido en la conversion de oxidos de nitrogeno. Revista Cubana de Física. 2020;**37**:108-116

[4] Glarborg P. Hidden interactions-trace species governing combustion and emissions. Proceedings of the Combustion Institute. 2007;**31**:77-98. DOI: 10.1016/j.proci.2006.08.119

[5] Abian M, Alzueta MU, Glarborg P. Formation of NO from N_2/O_2 mixtures in a flow reactor: Toward an accurate prediction of thermal NO. International Journal of Chemical Kinetics. 2015;**47**:518-532. DOI: 10.1002/kin.20929

[6] Hatakeyama K, Tanabe S, Hayashi Y, Matsumoto H. NO_X decomposition by discharge plasma reactor. Journal of Advanced Sciences. 2001;**13**:459-462. DOI: 10.2978/jsas.13.459

[7] Yoshida K, Rajanikanth BS, Okubo M. NO_X reduction and desorption studies under electric discharge plasma using a simulated gas mixture: A case study on the effect of corona electrodes. Plasma Science and Technology. 2009;**11**:327-333. DOI: 10.1088/1009-0630/11/3/15

[8] Balogh RM, Ionel I, Stepan D, Rabl HP, Pfaffinger A. NO_X reduction using selective catalytic reduction (SCR) system-α variation test. Termotehnica. 2011;**2**:32-42

[9] CreyghtonY, Pulsed Positive Corona Discharges: Fundamental Study and Application to Flue Gas Treatment [Thesis]. Technische University of Eindhoven Netherlands; 1994

[10] Dujko S, Ebert U, White RD, Petrović ZL. Boltzmann equation analysis of electron transport in a N_2/O_2 streamer discharge. Japanese Journal of Applied Physics. 2011;**50**(8S1):08JC01. DOI: 10.1143/JJAP.50.08JC01

[11] Batina J, Noël F, Lachaud S, Peyrous R, Loiseau JF. Hydrodynamical simulation of the electric wind in a cylindrical vessel with positive point-to-plane device. Journal of Physics D: Applied Physics. 2001;**34**:1510-1524. DOI: 10.1088/0022-3727/34/10/311

[12] Ono R, Oda T. Visualization of streamer channels and shock waves generated by positive pulsed corona discharge using laser Schlieren method. Japanese Journal of Applied Physics. 2004;**43**:321-237. DOI: 10.1143/JJAP.43.321

[13] Eichwald O, Yousfi M, Hennad A, Benabdessadok MD. Coupling of chemical kinetics, gas dynamics, and charged particle kinetics models for the analysis of NO reduction from flue gases. Journal of Applied Physics. 1997;**82**:4781-4794. DOI: 10.1063/1.366336

[14] Dorai R, Hassouni K, Kushner MJ. Interaction between soot particles and NO_X during dielectric barrier discharge

plasma remediation of simulated diesel exhaust. Journal of Applied Physics. 2000;**88**:6060-6071. DOI: 10.1063/1.1320004

[15] Loiseau JF, Batina J, Noël F, Peyrous R. Hydrodynamical simulation of the electric wind generated by successive streamers in a point-to-plane reactor. Journal of Physics D: Applied Physics. 2002;**35**:1020-1031. DOI: 10.1088/0022-3727/35/10/310

[16] Bouzar M, Ferouani AK, Lemerini M, Hocini AK. Zero-dimensional model description of the effect of NOx removal in $N_2/O_2/H_2O/CO_2$ mixtures in a nonuniform field. High Temperature Material Processes: An International Quarterly of High-technology Plasma Processes. 2017;**21**:225-237. DOI: 10.1615/HighTempMatProc.2018025371

[17] Chang JS. Physics and chemistry of plasma pollution control technology. Plasma Sources Science and Technology. 2008;**17**:045004. DOI: 10.1088/0963-0252/17/4/045004

[18] Ferouani AK, Lemerini M, Merad L, Houalef M. Numerical modelling point-to-plane of negative corona discharge in N_2 under non-uniform electric field. Plasma Science and Technology. 2015;**17**:469-474. DOI: 10.1088/1009-0630/17/6/06

[19] Ferouani AK, Lemerini, Belhour S. Numerical modelling of nitrogen thermal effects produced by the negative dc corona discharge. Plasma Science and Technology. 2010;**12**:208-211. DOI: 10.1088/1009-0630/12/2/15

[20] Flitti A, Pancheshnyi S. Gas heating in fast pulsed discharges in N_2–O_2 mixtures. The European Physical Journal-Applied Physics. 2009;**45**:21001. DOI: 10.1051/epjap/2009011

[21] Zhao L, Adamiak K. EHD flow in air produced by electric corona discharge in pin–plate configuration. Journal of Electrostatics. 2005;**63**:337-350. DOI: 10.1016/j.elstat.2004.06.003

[22] Ferouani AK, Lemerini M, Sahlaoui M, Askri S, Khaldi MFZ, Boumellah Y. The influence of radical N, O and O_3 in the reduction of nitrogen oxides in a corona discharge. Journal of Fundamental and Applied Sciences. 2021;**13**:931-941. DOI: 10.4314/jfas.v13i2.16

[23] Capitelli M, Ferreira CM, Gordiets BF, Osipov AI. Plasma Kinetics in Atmospheric Gases. Springer Science et Business Media; 2013. p. 31. DOI: 10.1007/978-3-662-04158-1

[24] Pancheshnyi S. Effective ionization rate in nitrogen–oxygen mixtures. Journal of Physics D: Applied Physics. 2013;**46**:155201. DOI: 10.1088/0022-3727/46/15/155201

[25] Zhao H, Lin H. Dielectric breakdown properties of N_2–O_2 mixtures by considering electron detachments from negative ions. Physics of Plasmas. 2016;**23**:073505. DOI: 10.1063/1.4956466

[26] Hoesl A, Haefliger P, Franck CM. Measurement of ionization, attachment, detachment and charge transfer rate coefficients in dry air around the critical electric field. Journal of Physics D: Applied Physics. 2017;**50**:485207. DOI: 10.1088/1361-6463/aa8faa

[27] Itikawa Y. Cross sections for electron collisions with nitrogen molecules. Journal of Physical and Chemical Reference Data. 2006;**35**:31-53. DOI: 10.1063/1.1937426

[28] Itikawa Y. Cross sections for electron collisions with oxygen molecules. Journal of Physical and Chemical Reference

Data. 2009;**38**:1-20. DOI: 10.1063/1.3025886

[29] Bekstein A, Yousfi M, Benhenni M, Ducasse O, Eichwald O. Drift and reactions of positive tetratomic ions in dry, atmospheric air: Their effects on the dynamics of primary and secondary streamers. Journal of Applied Physics. 2010;**107**:103308. DOI: 10.1063/1.3410798

[30] Stefanović I, Bibinov NK, Deryugin AA, Vinogradov IP, Napartovich AP, Wiesemann K. Kinetics of ozone and nitric oxides in dielectric barrier discharges in O_2/NOx and N_2/O_2/NOx mixtures. Plasma Sources Science and Technology. 2001;**10**:406-416. DOI: 10.1088/0963-0252/10/3/303

[31] Fan X, Kang S, Li J, Zhu T. Conversion of dilute nitrous oxide (N_2O) in N_2 and N_2–O_2 mixtures by plasma and plasma-catalytic processes. RSC Advances. 2018;**8**:26998-27007. DOI: 10.1039/C8RA05607B

[32] Haefliger P, Hösl A, Franck CM. Experimentally derived rate coefficients for electron ionization, attachment and detachment as well as ion conversion in pure O_2 and N_2–O_2 mixtures. Journal of Physics D: Applied Physics. 2018;**51**:355201. DOI: 10.1039/C8RA05607B

[33] Yamamoto T, Yang CL, Beltran MR, Kravets Z. Plasma-assisted chemical process for NO/sub x/control. IEEE Transactions on Industry Applications. 2000;**36**:923-927. DOI: 10.1109/28.845073

[34] Mei-Xiang P, Lin H, Shangguan WF, Huang Z. Simultaneous catalytic removal of NO_X and diesel PM over $La_{0.9}K_{0.1}CoO_3$ catalyst assisted by plasma. Journal of Environmental Sciences. 2005;**17**:220-223. DOI: 10.1109/28.845073

[35] Wang XQ, Chen W, Guo QP, Li Y, Lv GH, Sun XP, et al. Characteristics of NO_X removal combining dielectric barrier discharge plasma with selective catalytic reduction by C_2H_5OH. Journal of Applied Physics. 2009;**106**:013309. DOI: 10.1063/1.3160294

[36] Abedi-Varaki M, Ganjovi A, Shojaei F, Hassani Z. A model based on equations of kinetics to study nitrogen dioxide behavior within a plasma discharge reactor. Journal of Environmental Health Science and Engineering. 2015;**13**:69-78. DOI: 10.1186/s40201-015-0228-5

[37] Kossyi IA, Kostinsky AY, Matveyev AA, Silakov VP. Kinetic scheme of the non-equilibrium discharge in nitrogen-oxygen mixtures. Plasma Sources Science and Technology. 1992;**1**:207-220. DOI: 10.1088/0963-0252/1/3/011

[38] Atkinson R, Baulch DL, Cox RA, Hampson RF, Kerr JA, Rossi MJ, et al. Photochemical and heterogeneous data for atmospheric chemistry: Supplement V. IUPAC Subcommittee on gas Kinetic data evaluation for atmospheric chemistry. Journal of Physical and Chemical Reference Data. 1997;**26**:521-1011. DOI: 10.1063/1.556011

[39] Rosocha LA, Anderson GK, Bechtold LA, Coogan JJ, Heck HG, Kang M, et al. Non-Thermal Techniques for Pollution Control, Part B. Berlin: Springer-Verlag; 1993. p. 281. DOI: 10.1007/978-3-642-78476-7_21

[40] Eichwald O, Guntoro NA, Yousfi M, Benhenni M. Chemical kinetics with electrical and gas dynamics modelization for NOx removal in an air corona discharge. Journal of Physics D: Applied Physics. 2002;**35**:439-450. DOI: 10.1088/0022-3727/35/5/305

Chapter 4

Perspective Chapter: Numerical Simulation to Study Alfvén Waves

Bheem Singh Jatav

Abstract

This chapter presents the Alfvén waves self-interaction in approximately zero-β plasma (cold plasma), which is applicable to coronal heating. We consider thermal to magnetic pressure ratio $\beta \approx 0$ (cold plasma). When we consider magnetic field transverse perturbation in approximately zero-β plasma, the model dynamical equation of Alfvén wave self interaction turns out to be the modified Zakharov system equation. Numerical simulation has been carried out to study the effect of Alfvén wave self interaction and transverse perturbation, which results in the formation of magnetic field profile structures (coherent structures) and power spectrum for coronal heating and particle acceleration in space plasma. The investigated results reveals that the system of magnetic field profile structures reaches to quasi-steady state and the power spectral index approaches the $k^{-5/3}$, which is consistent with Kolmogorov scaling.

Keywords: Alfven wave, plasma, turbulence, numerical simulation, coronal heating

1. Introduction

Alfvén waves are low-frequency, electromagnetic waves in uniform magnetized plasma with a background magnetic field. It is well known that the mechanical effects of a magnetic field B_0 in a magnetized plasma are equivalent to an isotropic magnetic pressure $p_B = B_0^2/2\mu_0$, combined with a magnetic tension $T_B = B_0^2/\mu_0$ along the magnetic field lines. Analogy with the theory of stretched strings suggests that this tension may lead to a transverse wave propagating along the field lines, with the Alfven velocity $V_A = \sqrt{T_B/\rho} = B_0^2/\sqrt{\mu_0 \rho}$, where ρ is the mass density of magnetized plasma. These waves were first predicted in classic work by H. Alfvén [1] and can be derived from magnetrohydrodynamic (MHD) equation. These waves propagate parallel to ambient magnetic field. Alfvén waves were first experimentally demonstrated in laboratory by Lundquist (in mercury) [2] and Lehnert (in sodium) [3]. These waves plays very important role in space plasma. so we should understand the mechanism of Alfvén wave formation. The Waves of any kind in nature are driven by some restoring force which opposes displacements in the system. So the restoring force is responsible for the Alfven wave generation, which can be understood by two physical principles in conducting fluid.

i. Lenz's law applied to conducting Fluids- 'Electrical currents induced by the motion of a conducting fluid through a magnetic field give rise to electromagnetic forces acting to oppose that fluids motion.'

ii. Newton's second law for Fluids – 'A force applied to a fluids will result in a change in the momentum of the fluids proportional to the magnitude of the force and in the same direction.'

In general terms, the AWs are transient electromagnetic-hydromagnetic phenomenon occurring in magnetized plasma. These waves are traveling oscillations of ions and magnetic field. The ion mass density provides the inertia and the magnetic field lines tension provides the restoring force in magnetized plasma. The electron also oscillate in an Alfvén wave along the magnetic field lines, and they are responsible for the field-aligned current in Alfvén wave. Alfvén waves are electromagnetic wave, so they oscillation electric field. Inside Alfvén waves, electric and magnetic field oscillate perpendicular to the ambient magnetic field. Together with the electric and magnetic field yield a pointing vector that explains the energy flow of the wave. The oscillation of electric and magnetic field can be understood via an experiment [Davidson, 2001].

Let us look at a system of uniform magnetic field permeating a perfectly conducting fluid, which are uniformly flow initially to the magnetic field lines (**Figure 1**). The fluid flow will distort the magnetic field lines, so magnetic field lines become curved as shown in fig. (b). The curved of magnetic field lines produce a Lorentz force on the conducting fluid that opposes the next curvature as explained by

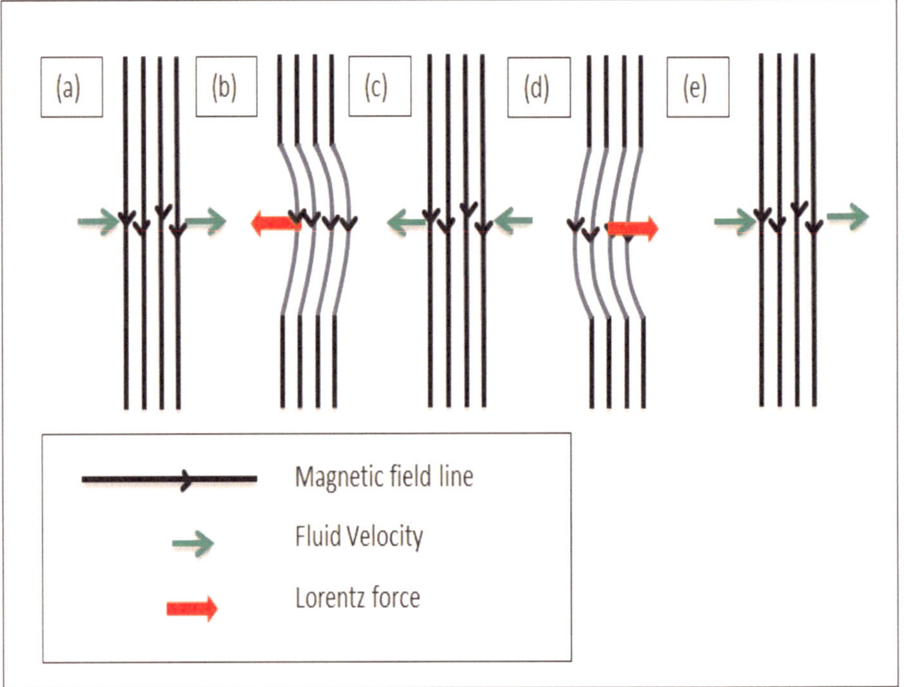

Figure 1.
Mechanism of Alfvén wave formation: (a) fluid velocity perpendicular to the uniform magnetic field (b) distorted magnetic lines giving rise to a Lorentz force that retards and eventually reverses the fluid velocity (c) field lines return to undisturbed position (d) reversed distortion and (e) repeats the cycle.

Lenz's Law. According to the Newton's second law, the Lorentz force changes the momentum of conducting fluid, which pushing it to minimize the magnetic field line distortion and restore the system in its equilibrium state. The restoring force provides the basis for the transverse oscillation of magnetic field in conducting fluid. As the curvature of magnetic field lines increases, so the Lorentz force becomes strong enough to chance the direction of the conducting fluid flow from initial direction. Now the magnetic field lines are pushed back to their undistorted state and the Lorentz force associated with their curvature weakens until the field lines becomes strength again. The sequence of flow of magnetic field lines distortion will repeat, but in opposite direction. This cycle will continue indefinitely, until the dissipation will apply.

The electromagnetic field oscillation generated in magnetic field lines by fluid flow and the elastic string when it plucked, having same mechanism. For both, the tension as a restoring force, magnetic tension in case of magnetic field lines and elastic tension in case of the string. The both waves resultant oscillations propagating in direction perpendicular to their displacement. The resultant oscillation generated by restoring force is the shear Alfven wave.

The low-frequency ($\omega < < \omega_{ci} = eB/cm_i$) and long wavelength ($\lambda_L > > \rho_i = v_{Ti}/\omega_{ci}$) perturbations apply in magnetized plasma, single- fluid MHD approach are more appropriate because $m_e \to 0$ and $n_e = n_i = n$. The low-frequency waves classified as the fast, slow and intermediate via phase speed of wave. The MHD analysis gives the following dispersion relation of Alfven waves are

$$v_A^2 = \frac{\omega^2}{k_{\parallel}^2} = \frac{B_o^2}{4\pi\rho} \tag{1}$$

The dispersion relation (Eq. (1)) of Alfven wave phase speed is independent of $k_{\perp}$ and perturbation of magnetic field perpendicular to the B_o and k.

$$\omega^2 = \frac{1}{2}k^2\left(v_A^2 + c_s^2\right) \pm \frac{k^2}{2}\sqrt{\left(v_A^2 + c_s^2\right)^2 - 4\frac{k_{\parallel}^2}{k^2}c_s^2 v_A^2} \tag{2}$$

The "+" and "- "sign in Eq. (2) leading to the fast magnetosonic wave $(\omega_f > \omega_i)$, where ω_i is the shear alfven frequency and slow magnetosonic wave $(\omega_s < \omega_i)$.

For low-β (Here $\beta = \left(2\pi n_0 T/B_0^2 = c_s^2/v_A^2 < < 1\right.$ is the thermal to magnetic pressure ratio) limit the dispersion relation reduce as

$$\omega_f^2 = k^2 v_A^2 \tag{3}$$

In the low-β limit the slow magnetosonic wave reduces to ion-acoustic wave and the dispersion relation can be written as

$$\omega^2 = k_{\parallel}^2 c_s^2 \tag{4}$$

The shear Alfvén wave characteristics depend on the ion gyroradius (ρ_s) or the electron skin depth (λ_e), which have determined as the dispersive Alfvén wave (Kletzing, 1994). The dispersion property of Alfvén waves is determined by the electron-β, where $\beta = \left(2\pi n_0 T_e/B_0^2 = (m_e/m_e)\left(2v_{Te}^2\right)\right.$ and $v_T e^2 = T_e/m_e$ is the electron thermal velocity. The Alfvén waves property depend upon the whether the Alfvén velocity is larger than the electron thermal speed ($v_{Te} < < v_A$, which is inirtial limit) or the smaller ($v_{Te} > > v_A$, which is the kinetic limit). Therefore, for kinetic limit the electrons moves fast are able to respond adiabatically to wave field. For low-β plasma ($\beta < < m_e/m_i$), the shear Alfvén wave call as the "Inertial Alfvén wave" whereas for intermediate-β plasma ($m_e/m_i < < \beta < < 1$), the shear Alfvén wave call as the "Kinetic Alfvén wave". The inertial limit is applicable for cold plasma (Aurora) and while the kinetic limit applicable to the hot plasma (Solar corona and solar wind) (Vincena et al., 2004).

The Alfvén waves have various application in solar physics. Alfvén waves have been proposed as energy transportation mechanism for corona [4, 5] to the accelerate solar wind [6]. The coronal heating is one of the important unsolved problem in solar physics from decades [7–10]. The coronal heating energy originated in convection zone to heat the coronal loops, but how this energy transported to coronal loops are not well understood. The one possibility of coronal heating is braided magnetic field dissipation in MHD models of coronal loops [11, 12]. Another one possibility to be the magnetic flux tubes, which generated by convective collapse [13, 14]. The transverse MHD waves can be formed in magnetic flux tube by interaction with granule scale convective flow [15]. These transverse MHD waves can propagate upward into the atmosphere and transport there energy in corona [16–19]. Alfvén wave are prominent because they can propagate at large distance in corona and transport there energy to coronal loops. Alfvén wave have been observed in corona [20–23], but the role of such waves are not well demonstrated. The 1.5D MHD simulation of torsional Alfvén waves explained, when the Alfvén waves are energized by random motion into photosphere, then the torsional Alfvén waves are sufficient to heat corona [24, 25]. The Alfvén wave nonlinearity generated by coupling of these waves to other MHD mode and leads to MHD shocks. Such type of shocks assume as coronal heating [26]. The 2.5D MHD model conclude that corona heated by fast and slow mode MHD shocks [27, 28].

The plasma exhibits a turbulent cascade over wide range of spectral scale [29]. The turbulent power spectrum consists several scales, in which various mechanism are possible in the transfer of energy from larger to smaller scales. The spectral power law behavior of the spectral indices steeper between power laws in ranges, leading to confirm different turbulent models [30–33]. MHD turbulence plays a prominent role to transfer wave energy in corona [34]. MHD turbulence is responsible for the transportation of energy from large scale to small scale where it is dissipated [35]. Compressible MHD turbulence proposed as a possible mechanism for wave energy dissipation to heat corona [36–38]. The wave energy dissipation process cause by the damping of MHD wave to heat corona by Alfvén waves [39, 40]. The interaction between two Alfvén wave leading to the wave energy cascade form source region to energy dissipation region through the MHD turbulence [41–44].

Alfvén waves are mixed mode having electric field parallel to the ambient magnetic field [45, 46]. Rosenthal and Bogdan [47, 48], numerically studied the wave propagation in a magneto-sphere and observed the positron generated both in fast and slow MHD wave to accelerate these wave to magnetized plasma. Their study explains the coupling of fast and slow mode waves, where the plasma beta unity. De Moortel et al. [49], studied the slow waves on 2D geometrical boundary with density variation. This author also shows the coupling between slow and fast and phase mixing of slow mode waves.

The coupling of fast wave and Alfvén wave has been studied by Parker [50] when the linear density gradient exist. When the nonlinear excitation exist,

Nakariakov et al. [51], investigated fast waves nonlinear excitation by Alfvén wave leading to phase mixing in inhomogeneous background density profile. Their numerical simulation study showed that the Alfvén waves not only heat via phase mixing dissipation method but also through nonlinear coupling of fast wave. Such type of phase mixing have been studied by various authors to investigate the coronal heating [52–55]. Thurgood and McLaughlim [56], studied the nonlinear damping mechanism for Alfvén wave at $\beta = 0$ (cold plasma), when the nonlinear ponderomotive force generated by the Alfvén waves can excite fast wave leading to the shock formation (coronal heating). The direct coupling between Alfvén and slow wave has been studied by Zaqarashvili et al. [57] when $\beta = 1$.

In this chapter we will study the Alfvén wave self-interaction with magnetic field transverse perturbation in cold plasma ($\beta \approx 0$). In this paper, we focus on the coronal heating by Alfvén wave self-interaction when consider the magnetic field transverse perturbation for cold plasma. This studies have carried out by numerical simulation method.

The content of chapter is organized as follows. In Section 2 the dynamical equation for Alfvén waves self interaction in cold plasma are derived. Section 3 consists the pseudo-spectral method. Section 4 explaines the numerical simulation and obtained results of our model equation. Finally, Section 5 deals with the conclusion of the work.

2. Model equation

We consider the Alfvén waves propagation in compressible, magnetically formed, and zero β plasma. Here we derive the dynamical equation of Alfvén wave self interaction with the help of the MHD set of equation, which can be written as [51].

$$\frac{\partial \rho}{\partial t} + \nabla.(\rho V) = 0 \tag{5}$$

$$\frac{\partial B}{\partial t} = \nabla \times (V \times B) \tag{6}$$

$$\rho\frac{\partial V}{\partial t} + \rho(V.\nabla)V = -\frac{(\nabla \times B) \times B}{4\pi} \tag{7}$$

$$\nabla.B = 0 \tag{8}$$

which connect the density ρ, magnetic field B and plasma velocity V.

We write Eqs. (5)–(8) in Cartesian coordinate (x, y, z) form. Assume the perturbation in the y direction is zero. The Cartesian form of Eqs. (5)–(8) are

$$\frac{\partial \rho}{\partial t} + \frac{\partial}{\partial x}(\rho V_x) + \frac{\partial}{\partial z}(\rho V_z) = 0, \tag{9}$$

$$\frac{\partial B_x}{\partial t} = -\frac{\partial}{\partial z}(V_z B_x - V_x B_z), \tag{10}$$

$$\frac{\partial B_y}{\partial t} = \frac{\partial}{\partial z}(V_x B_z - V_z B_y) - \frac{\partial}{\partial x}(V_x B_y - V_y B_x), \tag{11}$$

$$\frac{\partial B_z}{\partial t} = \frac{\partial}{\partial x}(V_z B_x - V_x B_z), \tag{12}$$

$$\rho\left[\frac{\partial V_x}{\partial t}+\left(V_x\frac{\partial V_x}{\partial x}+V_z\frac{\partial V_x}{\partial z}\right)\right]=-\frac{\left[-B_y\frac{\partial B_y}{\partial x}-B_z\left(\frac{\partial B_x}{\partial z}-\frac{\partial B_z}{\partial x}\right)\right]}{4\pi}, \tag{13}$$

$$\rho\left[\frac{\partial V_y}{\partial t}+\left(V_x\frac{\partial V_y}{\partial x}+V_z\frac{\partial V_y}{\partial z}\right)\right]=-\frac{\left[-B_z\frac{\partial B_y}{\partial z}-B_x\frac{\partial B_y}{\partial x}\right]}{4\pi}, \tag{14}$$

$$\rho\left[\frac{\partial V_z}{\partial t}+\left(V_x\frac{\partial V_z}{\partial x}+V_z\frac{\partial V_z}{\partial z}\right)\right]=-\frac{\left[B_y\frac{\partial B_y}{\partial z}+B_x\left(\frac{\partial B_x}{\partial z}-\frac{\partial B_z}{\partial x}\right)\right]}{4\pi}. \tag{15}$$

Further, we consider the equilibrium state of MHD system is in compressible plasma state of density ρ_0 with a uniform magnetic field $\vec{B}_0=B_0\hat{z}$. Initially, the medium is flow freely. The finite amplitude perturbation are taken as $\vec{V}(x,y,t)=0+V(x,y,z)$ for flow, $\vec{B}(x,y,t)=B_0+B(x,y,z)$ for magnetic field and $\rho=\rho_0+\rho_1(x,y,z)$ for density.

Then, the finite amplitude perturbation of inhomogeneous equilibrium state written by set of following equation

$$\frac{\partial\rho}{\partial t}+\frac{\partial}{\partial x}(\rho_0V_x)+\frac{\partial}{\partial z}(\rho_0V_z)=L_1, \tag{16}$$

$$\frac{\partial B_x}{\partial t}-B_0\frac{\partial V_x}{\partial z}=L_2, \tag{17}$$

$$\frac{\partial B_y}{\partial t}-B_0\frac{\partial V_y}{\partial z}=L_3, \tag{18}$$

$$\frac{\partial B_z}{\partial t}+B_0\frac{\partial V_x}{\partial x}=L_4, \tag{19}$$

$$\rho_0\frac{\partial V_x}{\partial t}-\frac{B_0\left(\frac{\partial B_x}{\partial z}-\frac{\partial B_z}{\partial x}\right)}{4\pi}=L_5, \tag{20}$$

$$\rho_0\frac{\partial V_y}{\partial t}-\frac{B_0\left(\frac{\partial B_y}{\partial z}\right)}{4\pi}=L_6, \tag{21}$$

$$\rho_0\frac{\partial V_z}{\partial t}=L_7. \tag{22}$$

Where the left-hand-side and right-hand–side terms govern the linear behavior and nonlinear behavior. The nonlinear terms L_1 to L_7, we get

$$L_1=-\frac{\partial}{\partial x}(\rho_1(x)V_x)-\frac{\partial}{\partial z}(\rho_1(x)V_z), \tag{23}$$

$$L_2=-\frac{\partial}{\partial z}(V_xB_x-V_xB_z), \tag{24}$$

$$L_3=-\frac{\partial}{\partial z}\left(V_zB_y-V_yB_z\right)+\frac{\partial}{\partial x}\left(V_yB_x-V_xB_y\right) \tag{25}$$

$$L_4=\frac{\partial}{\partial x}(V_zB_x-V_xB_z), \tag{26}$$

$$L_5 = -\rho_1(x)\left(V_x\frac{\partial V_x}{\partial x} + V_z\frac{\partial V_x}{\partial z}\right) - \rho_1(x)\frac{\partial V_x}{\partial t} - \rho_0\left(V_x\frac{\partial V_x}{\partial x} + V_z\frac{\partial V_x}{\partial z}\right) - \frac{1}{4\pi}B_y\frac{\partial B_y}{\partial x} + \frac{1}{4\pi}B_z\left(\frac{\partial B_x}{\partial z} - \frac{\partial B_z}{\partial x}\right), \quad (27)$$

$$L_6 = -\rho_1(x)\frac{\partial B_y}{\partial t} - \rho_1(x)\left(V_z\frac{\partial V_y}{\partial z} + V_x\frac{\partial V_y}{\partial x}\right) - \rho_0\left(V_x\frac{\partial B_y}{\partial x} - V_z\frac{\partial V_y}{\partial z}\right) + \frac{\left(B_x\frac{\partial B_y}{\partial x} + B_z\frac{\partial B_y}{\partial z}\right)}{4\pi}. \quad (28)$$

$$L_7 = -\rho_1(x)\frac{\partial V_z}{\partial t} - \rho_0\left(V_x\frac{\partial V_z}{\partial x} + V_z\frac{\partial V_z}{\partial z}\right) - \rho_1(x)\left(V_x\frac{\partial V_z}{\partial x} + V_z\frac{\partial V_z}{\partial z}\right) - \frac{1}{4\pi}B_y\frac{\partial B_y}{\partial z} - \frac{B_x\left(\frac{\partial B_x}{\partial z} - \frac{\partial B_z}{\partial x}\right)}{4\pi}. \quad (29)$$

Eqs. (16)–(22) consist the information about the propagation of Alfvén waves and fast magnetosonic waves in the Cartesian form. We used Eqs. (16)–(22) to obtain the nonlinear dynamical equation of Alfvén waves.

$$\frac{\partial^2 B_y}{\partial t^2} - {V_A}^2\frac{\partial^2 B_y}{\partial z^2} = \frac{1}{4\pi\rho_0}\left[\frac{\partial^2 B_y}{\partial x\partial t} + \frac{\partial^2 B_y}{\partial z\partial t}\right] + \frac{B_0}{4\pi}\frac{\partial^2 B_y}{\partial x\partial z} \quad (30)$$

where $V_A = B_0/\sqrt{4\pi\rho_0}$ is the Alfvén speed. The Eq. (30) describes Alfvén wave self interaction, which propagating in x-z plane, i.e. $\vec{k} = k_x\hat{x} + k_z\hat{z}$.

Consider the solution of Eq. (30) is as

$$B_y = \tilde{B}_y(x, z, t)e^{(k_{0x}x + k_{0z}z - \omega t)} \quad (31)$$

Substituting Eq. (31) into Eq. (30) and simplify equation for the case $\partial\tilde{B}_y < < k_{0_x}\tilde{B}_y$, here we get

$$(-2i\omega - ik_{0x} - ik_{0z})\frac{\partial\tilde{B}_y}{\partial t} - {V_A}^2\frac{\partial\tilde{B}_y}{\partial z} - i\left(\frac{B_0k_{0z}}{4\pi} - \frac{\omega}{4\pi\rho_0}\right)\frac{\partial\tilde{B}_y}{\partial x} - V_A^2\frac{\partial^2\tilde{B}_y}{\partial z^2} + \frac{B_0}{4\pi}k_{0x}k_{0z}\tilde{B}_y = 0 \quad (32)$$

Where ω is the Alfvén wave frequency.

The total magnetic field can be written in the form of field vector and transverse perturbation such that,

$$\tilde{B}_y = \tilde{B}_{y1} + \tilde{B}_{y2} \quad (33)$$

where $\tilde{B}_{y1}$ and $\tilde{B}_{y2}$ is the field vector and perturbation of Alfvén wave, then

$$\tilde{B}_{y2}\tilde{B}^*_{y2}\Big|_{x,z,t\neq 0} = \tilde{B}^2_{y21}\, exp\ [i(k_{0x}x + k_{0z}z - \omega t)] \quad (34)$$

The perturbed field form of Eq. (28) can be written as

$$(-2i\omega - ik_{0x} - ik_{0z})\frac{\partial\tilde{B}_{y21}}{\partial t} - V_A{}^2\frac{\partial\tilde{B}_{y21}}{\partial z} - i\left(\frac{B_0k_{0z}}{4\pi} - \frac{\omega}{4\pi\rho_0}\right)\frac{\partial\tilde{B}_{y21}}{\partial x} - V_A^2\frac{\partial^2\tilde{B}_{y21}}{\partial z^2}$$
$$+\frac{B_0}{4\pi}k_{0x}k_{0z}\left[\tilde{B}_{y21}\tilde{B}_{y21}^*\right] = 0. \tag{35}$$

Here we considered zero β plasma. Then the plasma heated by electron only i.e. $T_e \neq 0, T_i = 0$. The modified density fluctuation due to electron [58] can be written as

$$\frac{\delta n_s}{n_0} \approx -\xi\left|\tilde{B}_y\right|^2 \approx -\xi\left(\tilde{B}_y\tilde{B}_y^*\right) \approx \left(\tilde{B}_{y21}\tilde{B}_{y21}^*\right) \tag{36}$$

where $\xi = \left(1 + 8k_{0x}^2\lambda_e^2\right)/(48\pi n_0T_e)$.

From Eqs. (35) and (36), we get

$$(-2i\omega - ik_{0x} - ik_{0z})\frac{\partial\tilde{B}_{y21}}{\partial t} - V_A{}^2\frac{\partial\tilde{B}_{y21}}{\partial z} - i\left(\frac{B_0k_{0z}}{4\pi} - \frac{\omega}{4\pi\rho_0}\right)\frac{\partial\tilde{B}_{y21}}{\partial x} - V_A{}^2\frac{\partial^2\tilde{B}_{y21}}{\partial z^2}$$
$$+\frac{B_0}{4\pi}k_{0x}k_{0z}\left[-\xi\left(\tilde{B}_y\tilde{B}_y^*\right)\right] = 0. \tag{37}$$

The dimensionless form of (33) can be written as

$$-i\frac{\partial\tilde{B}_y}{\partial t} - 4\frac{\partial\tilde{B}_y}{\partial z} - i\frac{\partial\tilde{B}_y}{\partial x} - 4\frac{\partial^2\tilde{B}_y}{\partial z^2} - \left|\tilde{B}\right|_y^2\tilde{B}_y = 0 \tag{38}$$

We obtain the Eq. (37), which is the modified Zakharov system of equation [59]. The normalizing parameters for $\beta = 0$ plasma are

$$t_N \approx \frac{4\pi(2\omega + k_{0x} + k_{0z})}{B_0k_{0x}k_{0z}}, x_N \approx \left(\frac{\omega}{B_0\rho k_{0x}k_{0z}} + \frac{1}{k_{0x}}\right), z_N \approx \frac{4\pi V_A{}^2}{B_0k_{0x}k_{0z}}, B_N \approx \left[\frac{1 + 8k_{0x}^2\lambda_e^2}{48\pi n_0T_e}\right]^{-1/2}.$$

3. Numerical method

In this chapter, we have use pseudo-spectral method to solve partial differential Eq. (38).

3.1 Pseudo-spectral method

The pseudo-spectral method approaches the partial differential equation are solved pointwise in the physical space $(x - t)$. The Pseudo-Spectral methods have been successfully applied to the fluid dynamics (turbulence modeling, weather predictions) - non-linear waves - seismic modeling - MHD [60].

We have applied them to plasma turbulence simulations, and to the non-linear interaction of Alfven waves with plasmas. The pseudo-spectral method is applicable only on the spatial types of partial differential equations.

The general spatial partial differential equation is

$$Lu(x) = s(X), X\epsilon V \tag{39}$$

Boundary condition

$$f(u(y)) = 0, y\epsilon\partial V \tag{40}$$

Where L is the spatial differential operator $L = \partial/\partial x^2 + \partial/\partial y^2$. But here we need numerical soluation of $u^N(X)$.

The general procedure to find $u^N(X)$ are as follows.

Initially we choose a finite set of expanction function $\phi_j, j = 0, ..N - 1$, and expanding u^N by using these functions

$$u^N(X) = \sum_{j=0}^{N-1} \hat{u}_j\phi_j(X) \tag{41}$$

Further we choose a set of test function $\chi_n, k = 0,1,2, \dots \dots N - 1$ and demand that

$$(\chi_n, R) = 0, n = 0, 1. \dots N - 1 \tag{42}$$

Spectral methods' deals with the trial functions ϕ_n to form a basis for a certain space of smooth functions (Fourier expension).

The test function according to pseudo-spectral method can be written as

$$\chi_n(X) = \delta(X - X_n), \tag{43}$$

Where the X_n, $(n = 0, 1, \dots N - 1)$ are spatial point or the collection of points.

The smallness condition for the residual can be written as

$$0 - (\chi_n, R) = (\delta(X \quad X_n), R) = R(X_n) = Lu^N(X_n) - s(X_n) \tag{44}$$

$$\sum_{j=0}^{N-1} \hat{u}_j L\phi_j(X_n) - s(X_n) = 0, n = 0,1,2, \dots \dots N - 1. \tag{45}$$

Where N equations to determine the unknown N coefficients $\hat{u}_j$.

In this chapter applications, we assume periodic boundary condition and use in Fourier series as

$$\phi_j(X) = e^{-ik_jX} \tag{46}$$

Assume this case for 1-D only, then

$$\frac{\partial^2 u}{\partial z^2} = s(X) \tag{47}$$

The Fourier transform of Eq. (47) is

$$(-ik_z)^2\hat{u}(k_z) = \hat{s}(k_z) \tag{48}$$

We want to solve this numerically. Then our next aim is to find the expension coefficients $\hat{u}_j$ such that residual

$$\sum_{j=0}^{N-1} \hat{u}_j L\phi_j(X_N) - s(X_n) = 0, n = 0,1,2, \ldots, N-1. \tag{49}$$

or

$$\sum_{j=0}^{N-1} \hat{u}_j L e^{-ik_jX_n} - s(X_n) = 0, n = 0,1,2, \ldots, N-1. \tag{50}$$

vanishes.

If L considered as linear, then $Le^{-ik_KX_n} = h(k_k)e^{-ik_KX_n}$.

$$\sum_{j=0}^{N-1} \hat{u}_j h\left(k_j\right) e^{-ik_jX_n} - s(X_n) = 0, n = 0,1,2, \ldots, N-1. \tag{51}$$

For $1-D$ simplicity, we choose as

$$z_n = n\Delta, n = 0,1,2, \ldots N-1 \tag{52}$$

and

$$k_j = 2\pi j/(N\Delta), j = -N/2, \ldots., N/2. \tag{53}$$

where Δ is the spatial resolution and z_n and k_j are equi-spaced, then the condition on the residual becomes.

$$\sum_{j=0}^{N} \hat{u}_j h\left(k_j\right) e^{-2\pi ijn/N} - s(X_n) = 0, n = 0,1,2, \ldots, N-1. \tag{54}$$

Further we define the Discrete Fourier transform (DFT) as

$$\hat{u}_j = \frac{1}{\sqrt{N}} \sum_{n=0}^{N-1} u_n e^{-2\pi ijn/N}. \tag{55}$$

and inverse DFT written as

$$u_n = \frac{1}{\sqrt{N}} \sum_{j=-N/2}^{N/2} \hat{u}_j e^{-2\pi ijn/N}. \tag{56}$$

Then the DFT can be written as

$$\hat{u}_j h\left(k_j\right) - DFT[s(x_n)]\left(k_j\right) = 0. \tag{57}$$

Thus, we can manipulate our equations numerically with the DFT.

Finally, we calculate u_j by following steps.

(i) Give at time t at u_j. (ii) Calculate $u_n = DFT^{-1}(u_j)$. (iii) Multiply and store $T_n = u_n$ (iv) And move to F-space.

4. Numerical simulation results

It is very complicated to solve dynamical wave Eq. (38) analytically, hence the solution have carried out by numerical method [61, 62].

The numerical code have been written in FORTRAN using the technique of pseudo-spectral method with Fast Fourier Transform (FFT) for transverse (x-z direction) space integration with periodic length $(2\pi/\alpha_x) \times (2\pi/\alpha_z)$ at $\alpha_x, \alpha_z = 1$, implementing $2^8 \times 2^8$ grid points and a finite-difference method with a predictor-corrector scheme have used for the integration in the time domain step $dt = 5 \times 10^{-5}$. Before solving Eq. (38), we have applied our test code by analyzing the obtained result for solving the cubic nonlinear Schrödinger equation (CNSE). The precision of computation of nonlinear Schrödinger equation (NLSE) is constant $N = \sum_k |B_k|^2$. The conserved quantity was preserved up to the 10^{-5} during simulation.

The initial condition of simulation is

$$\tilde{B}_y(x, z, 0) = |a_0|[1 + 0.1cosh\,(\alpha_x x)][1 + 0.1cosh\,(\alpha_z z)] \tag{58}$$

Where $|a_0| = 1$ is the initial amplitude of Alfven wave.

Figure 2(a-d) depict the magnetic field intensity profile of Alfvén waves. Here, we consider four instant times namely t = 5,9,11,16. **Figure 2(a-b)** shows the propagation dynamics of Alfvén waves at t = 5,9 where the system reaches to quasi steady state. **Figure 2(c-d)** shows the formation of localized structures for Alfvén waves at t = 11, 16 and after few step the system reaches to its quasi steady state. The Alfvén waves spread there intensity peaks when the magnetic field intensity raised [63]. **Figure 2(a)** shows the Alfvén waves propagated without braking intensity at t = 5. **Figure 2(b)** shows the Alfvén waves propagated with breaking intensity. In **Figure 2(c)** shows the global spatial structures, where the perturbation takes energy from the interaction of Alfvén wave and form their complex structures (**Figure 2(d)**). At the initial time waves are propagated without breaking there intensity. When we take time advancement the waves intensity is broken and propagate small part of waves with different intensity peaks. For all instantaneous time the system reaches to quasi-steady state and formed more chaotic structures. These structures might play a crucial role in zero-β plasma for coronal heating and particle acceleration in space plasma.

Figure 3(a-d) depict the spectral contour plots of B_y in the (k_x, k_z) Fourier space at the same instants of time as in **Figure 2(a-d)** [64]. The most intense magnetic field structures are found in **Figure 3(c)**. But in **Figure 3(d)**, at t = 16, the more complex and chaotic state is found. At early time (t = 5) the complexity were decrease and at late time (t = 16) it increase. It is evident from contour plot that as we take time advancement the intensity and location of the magnetic field profile structures vary. And, with the advancement of time, the spectra in Fourier space become more chaotic and complex, leading to the energy transportation from larger scale to smaller scale (**Figures 3** and **4**).

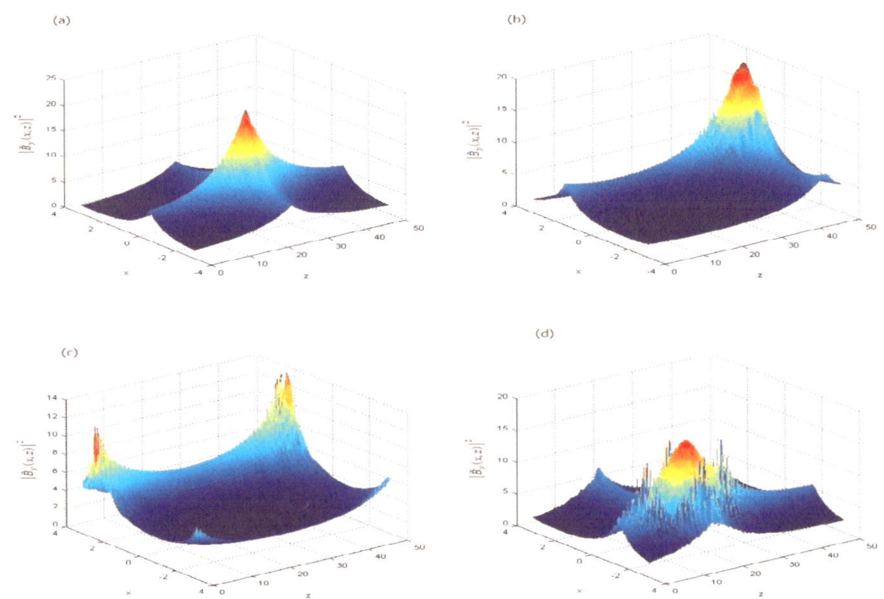

Figure 2.
The magnetic field intensity profile of AW for zero-β plasma at (a)t = 5, (b)t = 9, (c) t = 11, (d) t = 16.

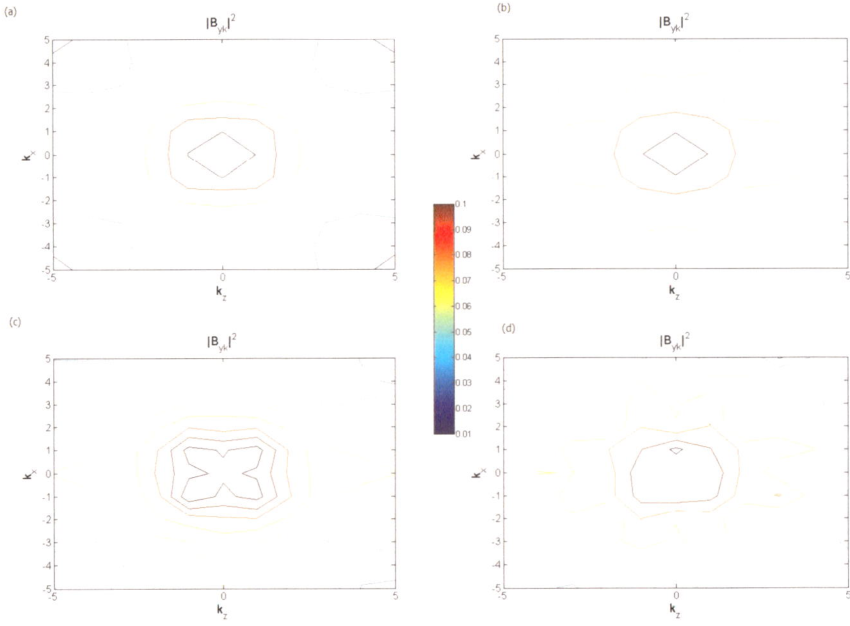

Figure 3.
The contour plot of $|B_{yk}|^2$ against Fourier modes of AW at (a) t = 5, (b) t = 9, (c) t = 11, (d) t = 16.

Next, we studied the power spectrum of magnetic field for Alfvén waves self interaction by plotting $|B_{yk}|^2$ against k at time t = 5, 9, 11, 16. We found the spectral index steepening range shows the Kolmogorov scaling at t = 5, i.e. $k^{-5/3}$, which is consistent with the particle acceleration in solar wind [41, 65, 66]. However, our

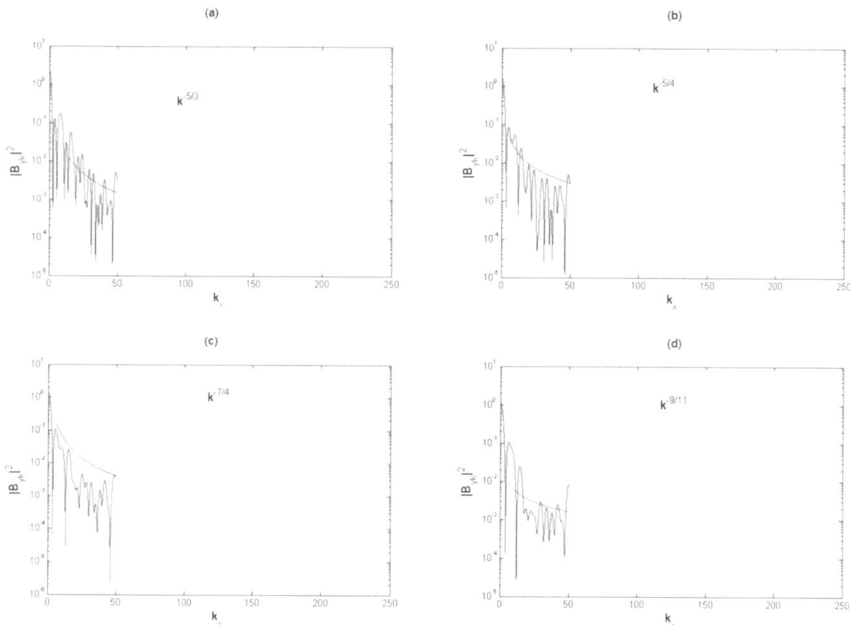

Figure 4.
The variation of $|B_{yk}|^2$ against k at (a)t = 5, (b)t = 9, (c) t = 11, (d) t = 16.

investigated results did not match the following reasons. The Alfvén waves are considered as Self-interaction in zero-β plasma. Also, the magnetic field transverse perturbation considered for Alfvén waves interaction in cold plasma. With the advancement of time ($t = 9{,}11{,}16$) power spectral index shows $k^{-5/4}, k^{-7/4}, k^{-9/4}$ scaling. The steepening of spectra with advancement of time govern to the coronal heating by Alfvén waves interaction. The consistent results observed by many authors, when the Alfvén waves interaction considered [34, 67, 68]. When the Alfvén waves interact nonlinearly in MHD turbulent plasma, then their spectra steeper the Kolmogorov like spectra ($k^{-5/3}$) studied by Goldreich and Sridhar [41, 59, 63, 64]; Goldstein et al. [66]. The strong turbulent studied by Schekochihin et al. [69] with slop $k^{-7/3}$ and observed particle asymptotic behavior in space plasma. The weak turbulent steepening of Alfvén waves were observed between kinetic and MHD range with slop -3 to -4 [70].

For application purpose, the typical values for solar Corona [71] are as follows: $B_0 \approx 10G, n_0 \approx 10^8, \beta \approx 10^{-2}, v_A \approx 2 \times 10^8, \lambda_e \approx 2 \times 10^3$. Using these values, one find, $\rho_s \approx 3.4 \times 10^9 cm, c_s \approx 2 \times 10^7 cm/s, \omega_{ci} \approx 5.8 \times 10^3$. For $k_{0x}\lambda_e \approx 0.98$, and $\frac{\omega_0}{\omega_{ce}} \approx 0.85$, one can calculate $k_{0x} \approx 4.9 \times 10^{-4} cm^{-1}, \omega_0 \approx 4.1 \times 10^3 sec^{-1}, k_{0z} \approx 2 \times 10^{-5} cm^{-1}$. The normalizing parameter are $x_n \approx 2 \times 10^3 cm, z_n \approx 5.1 \times 10^{24} cm, t_n \approx 1.1 \times 10^{12}, B_n \approx 1.5G$.

5. Conclusion

In this chapter, we have investigated the Alfvén waves self interaction in approximately zero-β plasma, when the magnetic field transverse perturbation considered. The numerical simulation of Eq. (38) carried out by pseudo-spectral method, which govern the dynamics of Alfvén waves interaction in cold plasma. The transverse perturbation leads to the steepening of power-spectrum which are $k^{-5/4}, k^{-7/4}, k^{-9/4}$ at time $t = 9,11,16$. The present chapter determine the Alfvén waves self interaction in approximately zero-β plasma for corona heating, when the transverse perturbation considered. We also investigate the relevance of proposed self-interaction to explain the observed power spectrum for the solar corona. Our observed results shows the spectrum steepens, which consequence to transport the energy from large scale to small scale. Observed results also revel that the transverse magnetic field profile structures (coherent structures) are break at highest intensity lead to cascade energy. Hence the transverse perturbation affect the coherent structures and power spectrum, govern to coronal heating and particle acceleration in space plasma.

Author details

Bheem Singh Jatav
Jaipur National University, Jaipur, Rajasthan, India

*Address all correspondence to: bheemsinghj@gmail.com

IntechOpen

© 2023 The Author(s). Licensee IntechOpen. This chapter is distributed under the terms of the Creative Commons Attribution License (http://creativecommons.org/licenses/by/3.0), which permits unrestricted use, distribution, and reproduction in any medium, provided the original work is properly cited.

References

[1] Alfven H. Existence of Electromagnetic-Hydrodynamic Waves. Nature. 1942;**150**:405

[2] Lundquist S. Experimental Investigations of Magneto-Hydrodynamic Waves. Physics Review. 1949;**76**:1805

[3] Lehnert B. Magneto-Hydrodynamic Waves in Liquid Sodium. Physics Review. 1954;**94**:815

[4] Kudoh T, Shibata K. Alfvén Wave Model of Spicules and Coronal Heating. The Astrophysical Journal. 1999;**512**:493

[5] van Ballegooijen AA, Asgari-Targhi M, Cranmer SR, DeLuca EE. Heating of the Solar Chromosphere and Corona by Alfvén Wave Turbulence. The Astrophysical Journal. 2011;**736**:3

[6] Klimchuk JA. On Solving the Coronal Heating Problem. Solar Physics. 2006; **234**:41

[7] Zirker JB. Coronal heating. Solar Physics. 1993;**148**:43

[8] Aschwanden MJ. Physics of the Solar Corona. An Introduction with Problems and Solutions. 2nd edn. Berlin: Springer; 2005

[9] De Moortel I, Browning P. Recent advances in coronal heating. Philosophical Transactions of the Royal Society. 2015;**373**:20140269

[10] van Ballegooijen AA, Asgari-Targhi M, Voss A. The Heating of Solar Coronal Loops by Alfvén Wave Turbulence. The Astrophysical Journal. 2017;**849**:46

[11] Rappazzo AF, Velli M, Einaudi G, Dahlburg RB. Coronal Heating, Weak MHD Turbulence, and Scaling Laws. The Astrophysical Journal. 2007;**657**:L47

[12] Dahlburg RB, Einaudi G, Taylor BD, Ugarte-Urra I, Warren HP, Rappazzo AF, et al. Observational Signatures of Coronal Loop Heating and Cooling Driven by Footpoint Shuffling. The Astrophysical Journal. 2016;**817**:47

[13] Spruit HC. Convective collapse of flux tubes. Solar Physics. 1979;**61**:363

[14] Fischer CE, de Wijn AG, Centeno R, Liter BW, Keller CU. Statistics of convective collapse events in the photosphere and chromosphere observed with the Hinode SOT. Astronomy & Astrophysics. 2009;**504**:583

[15] Mumford SJ, Fedun V, Erdélyi R. Generation of Magnetohydrodynamic Waves in Low Solar Atmospheric Flux Tubes by Photospheric Motions. The Astrophysical Journal. 2015;**799**:6

[16] Alfvén H, Lindblad B. Magneto hydrodynamic waves, and the heating of the solar corona. Monthly Notices of the Royal Astronomical Society. 1947;**107**: 211-219

[17] Hollweg JV. Alfvén waves in the solar atmosphere. Solar Physics. 1978;**56**:305

[18] Parnell CE, De Moortel I. A contemporary view of coronal heating. Philosophical Transactions of the Royal Society A. 2012;**370**:3217

[19] Cranmer SR, Asgari-Targhi M, Miralles MP, Raymond JC, Strachan L, Tian H, et al. The role of turbulence in coronal heating and solar wind expansion. Philosophical Transactions of the Royal Society A. 2015;**373**:20140148

[20] Tomczyk S, McIntosh SW, Keil SL, Judge PG, Schad T, Seeley DH, et al. Alfven waves in the solar corona. Science. 2007;**317**:1192

[21] Tomczyk S, McIntosh SW. Time-distance seismology of the solar corona with comp. The Astrophysical Journal. 2009;**697**:1384

[22] Threlfall J, De Moortel I, McIntosh SW, Bethge C. First comparison of wave observations from CoMP and AIA/SDO. Astronomy & Astrophysics. 2013;**556**:A124

[23] Morton RJ, Tomczyk S, Pinto R. Investigating Alfvénic wave propagation in coronal open-field regions. Nature Communications. 2015;**6**:7813

[24] Hollweg J. Alfvenically driven slow shocks in the solar chromosphere and corona. The Astrophysical Journal. 1992; **389**:731

[25] Kudoh T, Shibata K. Alfvén Wave Model of Spicules and Coronal Heating. The Astrophysical Journal. 1999;**514**:493

[26] Moriyasu S, Kudoh T, Yokoyama T, Shibata K. The Nonlinear Alfvén Wave Model for Solar Coronal Heating and Nanoflares. The Astrophysical Journal. 2004;**601**:L107

[27] Matsumoto T, Suzuki TK. Connecting the sun and the solar wind: the first 2.5-dimensional self-consistent mhd simulation under the alfven wave scenario. The Astrophysical Journal. 2012;**749**:8

[28] Matsumoto T, Suzuki TK. Connecting the Sun and the solar wind: The self-consistent transition of heating mechanisms. Monthly Notices of the Royal Astronomical Society. 2014;**440**:971

[29] Chen CHK. Recent progress in astrophysical plasma turbulence from solar wind observations. Journal of Plasma Physics. 2016;**82**:535820602

[30] Matthaeus WH, Goldstein ML. Measurement of the rugged invariants of magnetohydrodynamic turbulence in the solar wind. Journal of Geophysical Research. 1982;**87**:6011

[31] Horbury TS, Forman M, Oughton S. Anisotropic Scaling of Magnetohydrodynamic Turbulence. Physical Review Letters. 2008;**101**:175005

[32] Alexandrova O, Saur J, Lacombe C, Mangeney A, Mitchell J, Schwartz SJ, et al. Universality of Solar-Wind Turbulent Spectrum from MHD to Electron Scales. Physical Review Letters. 2009;**103**:165003

[33] Howes GG, Tenbarge JM, Dorland W, Quataert E, Schekochihin AA, Numata R, et al. Gyrokinetic Simulations of Solar Wind Turbulence from Ion to Electron Scales. Physical Review Letters. 2011;**107**:035004

[34] Matthaeus WH, Zank GP, Oughton S, Mullan DJ, Dmitruk P. Coronal Heating by Magnetohydrodynamic Turbulence Driven by Reflected Low-Frequency Waves. The Astrophysical Journal. 1999; **523**:L93

[35] Cranmer SR, van Ballegooijen AA, Edgar RJ. Self-consistent Coronal Heating and Solar Wind Acceleration from Anisotropic Magnetohydrodynamic Turbulence. The Astrophysical Journal. 2007;**171**:520

[36] Pettini M, Nocera L, Vulpiani A. Compressible MHD turbulence: An efficient mechanism to heat stellar coronae. In: Buchler JR, Perdang JM, Spiegel EA, editors. Chaos in Astrophysics. NATO ASI Series (Series C: Mathematical and Physical Sciences). Vol. 161. Dordrecht: Springer; 1985

[37] Hollweg J. Transition region, corona, and solar wind in coronal holes. Journal

of Geophysical Research: Space Physics. 1986;**91**:4111

[38] Kumar N, Kumar P, Singh S. Coronal heating by MHD waves. Astronomy & Astrophysics. 2006;**453**:1067

[39] Braginskii SI. Transport Processes in Plasma. Reviews of Plasma Physics. 1965;**1**:205

[40] Osterbrock DE. The Heating of the Solar Chromosphere, Plages, and Corona by Magnetohydrodynamic Waves. The Astrophysical Journal. 1961;**134**:347

[41] Goldreich P, Sridhar S. Toward a Theory of Interstellar Turbulence. II. Strong Alfvenic Turbulence. The Astrophysical Journal. 1995;**438**:763

[42] Tu C-Y, Pu Z-Y, Wei F-S.The power spectrum of interplanetary Alfvénic fluctuations: Derivation of the governing equation and its solution. Journal of Geophysical Research. 1984;**89**:9695

[43] Ng CS, Bhattacharjee A. Interaction of Shear-Alfven Wave Packets: Implication for Weak Magnetohydrodynamic Turbulence in Astrophysical Plasmas. The Astrophysical Journal. 1996;**465**:845

[44] Cranmer SR, van Ballegooijen AA. On the Generation, Propagation, and Reflection of Alfvén Waves from the Solar Photosphere to the Distant Heliosphere. The Astrophysical Journal. 2003;**594**:573

[45] Stefant RJ. Alfvén Wave Damping from Finite Gyroradius Coupling to the Ion Acoustic Mode. Physics of Fluids. 1970;**13**:440

[46] Lysak RL, Lotko W. On the kinetic dispersion relation for shear Alfvén waves. Journal of Geophysical Research. 1996;**376**:355

[47] Rosenthal CS, Bogdan TJ, Carlsson M, Dorch SBF, Hansteen V, McIntosh SW, et al. Waves in the Magnetized Solar Atmosphere. I. Basic Processes and Internetwork Oscillations. The Astrophysical Journal. 2002;**564**:508

[48] Bogdan TJ, Hansteen MCV, McMurry A, Rosenthal CS, Johnson M, Petty-Powell S, et al. Waves in the Magnetized Solar Atmosphere. II. Waves from Localized Sources in Magnetic Flux Concentrations. Nordlund, A. 2003; **599**:626

[49] De Moortel I, Hood AW, Gerrard CL, Brooks SJ. The damping of slow MHD waves in solar coronal magnetic fields. Astronomy & Astrophysics. 2004;**425**:741

[50] Parker EN. The Phase Mixing of Alfven Waves, Coordinated Modes, and Coronal Heating. The Astrophysical Journal. 1991;**376**:355

[51] Nakariakov VM, Roberts B, Murawski K. Alfvén Wave Phase Mixing as a Source of Fast Magnetosonic Waves. Solar Physics. 1997;**175**:93

[52] Heyvaerts J, Priest ER. Coronal heating by phase-mixed shear Alfven waves, Astron. Astronomy and Astrophysics. 1983;**117**:220

[53] Narain U, Ulmschneider P. Chromospheric and coronal heating mechanisms. Space Science Reviews. 1990;**54**:377

[54] Malara F, Primavera L, Veltri P. Formation of Small Scales via Alfven Wave Propagation in Compressible Nonuniform Media. The Astrophysical Journal. 1996;**459**:347

[55] Ireland J. Wave heating of the solar corona and SoHO. Annales de Geophysique. 1996;**14**:485

[56] Thurgood JO, McLaughlin JA. On Ponderomotive Effects Induced by Alfvén Waves in Inhomogeneous 2.5D MHD Plasmas. Solar Physics. 2013; **288**:205

[57] Zaqarashvili TV, Oliver R, Ballester JL. Spectral line width decrease in the solar corona: resonant energy conversion from Alfvén to acoustic waves. Astronomy & Astrophysics. 2006;**456**:L13

[58] Shukla PK, Stenflo L. Plasma density cavitation due to inertial Alfvén wave heating, Phys. Physics of Plasmas. 1999; **6**:4120

[59] Sharma RP, Singh HD. Journal of Astrophysics and Astronomy. 2008;**29**: 239-242

[60] Sandberg I, Isliker H, Pavlenko V, Hizanidis K, Vlahos L. Large-scale flows and coherent structure phenomena in flute turbulence. Physics of Plasmas. 2005;**12**:032503

[61] Singh HD, Jatav BS. Coherent structures and spectral shapes of kinetic Alfvén wave turbulence in solar wind at 1 AU. Research in Astronomy and Astrophysics. 2019;**19**(7):93

[62] Singh HD, Jatav BS. Anisotropic turbulence of kinetic Alfvén waves and heating in solar corona. Research in Astronomy and Astrophysics. 2019; **19**(12):185

[63] Sharma RP, Singh HD, Malik M, Journal of Geo-physical Research (Space Physics). 2006;**111**A12108

[64] Singh HD, Sharma RP. Solar Physics. 2007;**243**:219-229

[65] Goldstein ML, Roberts DA, Deane AE, Ghosh S, Wong HK. Numerical simulation of Alfvénic turbulence in the solar wind. Journal of Geophysical Research. 1999;**104**(A7): 14437-14451

[66] Sahraoui F, Goldstein ML, Robert P, Khotyaintsev YV. Evidence of a Cascade and Dissipation of Solar-Wind Turbulence at the Electron Gyroscale. Physical Review Letters. 2009;**102**: 231102

[67] Wentzel DG. Coronal heating by Alfvén waves. Solar Physics. 1974;**39**: 129-140

[68] Zanna LD, Velli M. Parametric decay of circularly polarized Alfvén waves: Multidimensional simulations in periodic and open domains. Astronomy & Astrophysics (A&A). 2001;**367**:705

[69] Schekochihin AA, Cowley SC, Dorland W, Hammett GW, Howes GG, Plunk GG, et al. Gyrokinetic turbulence: A nonlinear route to dissipation through phase space. Plasma Physics and Controlled Fusion. 2008;**50**:124024

[70] Voitenko Y, De Keyser J. Turbulent spectra and spectral kinks in the transition range from MHD to kinetic Alfvén turbulence. Nonlinear Processes in Geophysics. 2011;**18**:587

[71] Stasiewicz K. Heating of the Solar Corona by Dissipative Alfvén Solitons. Physical Review Letters. 2006;**96**: 175003.

Chapter 5

Unstable Electrostatic Waves Associated with Density and Temperature Gradients in an Inhomogeneous Plasma

Banashree Saikia and Paramananda Deka

Abstract

A study is carried out on the amplification of electrostatic Bernstein waves in the presence of drift wave turbulence in an inhomogeneous plasma. We have considered the Vlasov-Poisson system of equations for the interaction among the waves. In this investigation, we have considered a particle distribution model in which an external force due to density and temperature gradients present in the system is incorporated. The resonant mode of drift wave turbulence interacts with plasma particles through resonant interactions. These accelerated particles transfer their energy to Bernstein waves through a modulated field. This nonlinear wave interaction process is based on weak turbulence theory and is known as the Plasma maser effect. In this process, we have evaluated the fluctuating parts of the distribution function owing to the presence of turbulent resonant waves, the modulation field of interacting waves, and the non-resonant Bernstein waves by integrating along the unperturbed particles orbit from a set of linearized Vlasov equations. The nonlinear dispersion relation for Bernstein waves is presented to analyse the influence of gradient parameters.

Keywords: nonlinear wave particle interaction, Bernstein wave, density gradient, temperature gradient, drift wave turbulence

1. Introduction

The Bernstein waves were first observed and studied in the 1950s by an American physicist named Ira Bernstein, who made significant contributions to the field of plasma physics. These waves play a crucial role in understanding various phenomena in astrophysics and fusion research. In the last decade, the Bernstein waves have been studied in a wide range of linear and nonlinear phenomena, and they still represent a major area of interest in fundamental research. Satellite observations of space plasmas have been interpreted using the findings of such investigations. Additionally, strategies for heating plasmas and diagnosing plasma behaviour have been developed in this regard.

Observations of Bernstein waves have been made both in laboratory experiments and in measurements of magnetospheric plasma carried out by spacecrafts. Stix [1] identified that there were two kinds of Bernstein waves: one was electron mode and the other was ion mode. It is interesting to note that electron Bernstein waves have been observed in both Earth's and Jupiter's magnetospheres. Kurth et al. [2] and Stix [1] were the first to observe these waves in Earth's magnetosphere, while Grimaldo et al. [3] more recently observed them in Earth's plasma. Experimentally, the electron Bernstein wave has been studied in tokamak plasma, especially in the spherical tokamak [4]. It seems that current research on electron cyclotron wave heating and current drive in magnetically confined thermonuclear fusion is focused on conventional aspect ratio tokamaks, with particular emphasis on ITER (International Thermonuclear Experimental Reactor). Electrostatic Bernstein waves have been studied in depth in various contexts, including the solar atmosphere, the auroral region, and through analytical and numerical methods. Additionally, the generation of electrostatic Bernstein modes through nonlinear wave-particle interaction in the presence of drift wave turbulence has been explored by several authors [5–8]. Deka [9] investigated the amplification of the electrostatic Bernstein mode in the presence of ion-acoustic turbulence, while Singh [10] studied its generation in the presence of kinetic Alfven wave turbulence and a density gradient in the medium.

Singh and Deka [11] investigated the amplification of the electrostatic Bernstein mode in the presence of drift wave turbulence, and Singh [8] investigated the amplification in the presence of a density gradient in the media. In order to study Bernstein-ring instability, Yoon et al. [12] considered a more accurate electron distribution function. They indicate that the absence of a strong perpendicular velocity gradient in the model distribution suggests that the electrostatic Bernstein waves cannot be excited. In the Saturnian magnetosphere, Henning et al. [13] studied the electrostatic Bernstein waves in plasmas with dual kappa electron distributions.

We have investigated the instability of Bernstein waves propagating perpendicular to the external magnetic field under the influence of two types of waves: [14] one is the resonant low-frequency drift wave, satisfying the Cherenkov resonance condition, and the other is a non-resonant high frequency electromagnetic extraordinary mode wave, satisfying neither the Cherenkov resonance condition nor the nonlinear scattering conditions. They have found that when both waves are present, a new maser-like emission occurs, which is called the plasma maser effect [15–17].

We demonstrate how the above mentioned process causes energy upconversion from the resonant drift mode to the non-resonant Bernstein mode wave. It is illustrated that the nonlinear dielectric constant is made up of two parts: direct coupling terms and polarisation mode coupling terms.

The chapter is structured as follows: In Section 2, the geometrical framework and mathematical formulation of our model is outlined, providing a solid foundation for the subsequent analysis. Following this, Section 3 presents the nonlinear dispersion relation of the Bernstein wave. The results and discussions are given in Section 4. Finally, in Section 5, we delve into a conclusion of the analytical investigations conducted in this chapter and explore potential applications that can arise from them.

2. Basic framework of the problem

2.1 Mathematical formulation

In this chapter, we have considered an inhomogeneous plasma that supports drift motions and the turbulence that results from the drift waves. For this framework, we consider a particle distribution function [18] involving density and temperature gradient parameters connected to the external force F. The gradients of density and temperature are taken along the y-direction (**Figure 1**).

$$f_j(T_j, y, \boldsymbol{v}) = \left(\frac{m}{2\pi T_{0j}}\right)^{\frac{3}{2}} \left[1 + \lambda\left(y + \frac{v_x}{\Omega_j}\right)\right] exp\left[-\left(\frac{m\boldsymbol{v}^2}{2\boldsymbol{T}_{0j}} - \frac{\boldsymbol{F}y}{\boldsymbol{T}_{0j}}\right)\right] \tag{1}$$

Where

$$\lambda = \left[\left(\frac{\partial}{\partial T_j}\frac{dT_j}{dy}\right) - \left(\frac{1}{f_j}\frac{df_j}{dy}\right)\right]_{y=0} - \frac{\boldsymbol{F}}{T_{0j}} \tag{2}$$

The interaction of high-frequency electrostatic Bernstein wave with drift wave turbulence is governed by Vlasov-Poisson system of equations.

$$\left[\frac{\partial}{\partial t} + \boldsymbol{v}.\frac{\partial}{\partial \boldsymbol{r}} + \left\{\frac{e}{m_j}\left(\boldsymbol{E} + \frac{\boldsymbol{v}\times\boldsymbol{B}_0}{c}\right) - \frac{\boldsymbol{F}}{m_j}\right\}.\frac{\partial}{\partial \boldsymbol{v}}\right] F_{0j}(\boldsymbol{r}, \boldsymbol{v}, t) = 0 \tag{3}$$

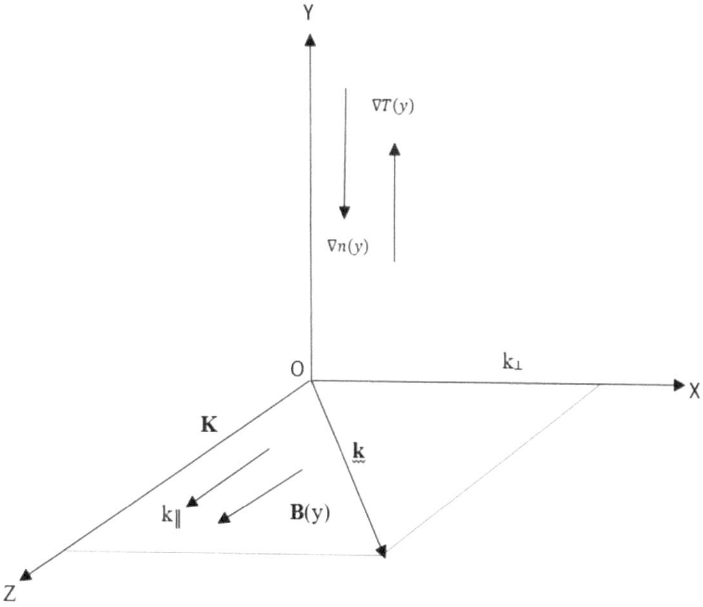

Figure 1.
Geometrical framework of the model: $\boldsymbol{K}$ = $(0, 0, K_{\parallel})$ *indicates the propagation vector of the Bernstein wave.* $\boldsymbol{k} = (k_{\perp}, 0, k_{\parallel})$ *indicates the propagation vector of the drift wave.* $\nabla n(y)$ *indicates the density gradient along the negative y-direction.* $\boldsymbol{B}(y)$ *indicates the magnetic field along z-direction.* $\nabla T(y)$ *indicates the magnetic field gradient along the +ve y-direction.*

$$\nabla.\boldsymbol{E} = -4en_j\pi\int f_{0j}(\boldsymbol{r},\boldsymbol{v},t)d\boldsymbol{v} \tag{4}$$

The linear response theory [18] of a turbulent plasma describes the unperturbed electron distribution function and the unperturbed electric field as:

$$F_{0j} = f_{0j} + \varepsilon f_{1j} + \varepsilon^2 f_{2j} \tag{5}$$

$$\vec{E}_{0l} = \varepsilon\vec{E}_l + \varepsilon^2\vec{E}_2 \tag{6}$$

where f_{0j} is the space and time averaged component of the distribution function, f_{1j} and f_{2j} are fluctuating parts due to the drift wave turbulence of the distribution function, and ε is an extremely small parameter associated with the low frequency drift wave turbulence.

To the order of ε, we obtain from Eq. (3)

$$\left[\frac{\partial}{\partial t} + \boldsymbol{v}.\frac{\partial}{\partial \boldsymbol{r}} + \left\{\frac{e}{m_j}\left(\varepsilon\boldsymbol{E}_l + \varepsilon^2\boldsymbol{E}_2 + \frac{\boldsymbol{v}\times\boldsymbol{B}_0}{c}\right) - \frac{\boldsymbol{F}}{m_j}\right\}.\frac{\partial}{\partial\boldsymbol{v}}\right]f_{1j}(\boldsymbol{r},\boldsymbol{v},t) = \frac{e_j}{m_j}\left(\boldsymbol{E}_l.\frac{\partial}{\partial\boldsymbol{v}}f_{0j}\right) \tag{7}$$

By employing the Fourier transform in the several quantities according to

$$A(\boldsymbol{r},\boldsymbol{v},t) = \sum_{k,\,\omega} A(\boldsymbol{k},\omega,\boldsymbol{v})\exp[i(\boldsymbol{k}.\boldsymbol{r} - \omega t)] \tag{8}$$

We perturb the quasi-steady state with the test Bernstein wave field $\mu\delta\overrightarrow{E_h}$ and wave vector $K = (0, 0, K_\parallel)$. The overall perturbed electric field as well as the electron distribution function as a result of this perturbation are as follows:

$$\delta\boldsymbol{E} = \mu'\delta\boldsymbol{E}_h + \mu'\varepsilon\delta\boldsymbol{E}_{lh} + \mu'\varepsilon^2\Delta\boldsymbol{E} \tag{9}$$

$$\delta f_j = \mu'\delta f_h + \mu'\varepsilon\delta f_{lh} + \mu'\varepsilon^2\Delta f \tag{10}$$

where μ' is another small parameter ($\mu' \ll \varepsilon$) and $\delta\boldsymbol{E}_{lh}$, $\Delta\boldsymbol{E}$, δf_{lh} and Δf are the mixed mode perturbation parameters. The omission of second order field quantities is justified by random phase approximations. To the order of μ, $\mu\varepsilon$ and $\mu\varepsilon^2$ we get

$$P\delta f_h = \frac{e_j}{m_j}\left(\boldsymbol{E}_h + \frac{\boldsymbol{v}\times\delta\boldsymbol{B}_h}{c}\right).\frac{\partial f_{0j}}{\partial\boldsymbol{v}} \tag{11}$$

$$P\delta f_{lh} = \frac{e}{m_j}\delta\boldsymbol{E}_h.\frac{\partial f_{1j}}{\partial v} + \frac{e}{m_j}\frac{\boldsymbol{v}\times\delta\boldsymbol{B}_h}{c}.\frac{\partial f_{1j}}{\partial\boldsymbol{v}} + \frac{e}{m_j}\delta\boldsymbol{E}_{lh}.\frac{\partial f_{0j}}{\partial\boldsymbol{v}} + \frac{e}{m_j}\delta\boldsymbol{E}_l.\frac{\partial f_h}{\partial\boldsymbol{v}} + \frac{e}{m_j}\frac{\boldsymbol{v}\times\delta\boldsymbol{B}_{lh}}{c}.\frac{\partial f_{0j}}{\partial\boldsymbol{v}} \tag{12}$$

$$P\Delta f = \frac{e}{m_j}\left[\delta\boldsymbol{E}_{lh}.\frac{\partial f_{1j}}{\partial\boldsymbol{v}} + \frac{\boldsymbol{v}\times\delta\boldsymbol{B}_{lh}}{c}.\frac{\partial f_{1j}}{\partial\boldsymbol{v}} + \boldsymbol{E}_l.\frac{\partial\delta f_{lh}}{\partial\boldsymbol{v}}\right] \tag{13}$$

where $P = \left[\frac{\partial}{\partial t} + \boldsymbol{v}.\frac{\partial}{\partial\boldsymbol{r}} - \left\{\frac{e}{m_j}\left(\frac{\boldsymbol{v}\times\boldsymbol{B}_0}{c}\right) - \frac{F}{m_j}\right\}.\frac{\partial}{\partial\boldsymbol{v}.}\right]$

Eqs. (11), (12) and (13) are the fundamental equations for the induced Bremsstrahlung interaction, which results from the electron acceleration driven by nonlinear forces.

3. Non-linear dispersion relation of Bernstein wave

The fluctuating parts of the low frequency turbulent field f_{1j} are now obtained from Eq. (8)

$$f_{1j} = \int_{-\infty}^{0} \frac{e}{m} \vec{E_l} . \frac{\partial f_{0j}}{\partial \vec{v}} \exp\left[i.\left\{\vec{k}.\left(\vec{r}' - \vec{r}\right) - \omega\tau\right\}\right] d\tau \tag{14}$$

Which takes the form

$$f_{1j} = \frac{ie}{m} \frac{E_{l\|}}{\omega - k_{\|} v_{\|} + i.0^{+}} \frac{\partial f_{0j}}{\partial v_{\|}} \tag{15}$$

Where $i0^{+}$ is a small imaginary part of ω, $\vec{k}$ is the wave number of the drift wave, and $\|$ means parallel to the magnetic field.

We obtain other fluctuating parts of the perturbed distribution function from Eqs. (11) to (13) as follows:

$$\begin{aligned}
\delta f_h &= \frac{e}{m} \int_{-\infty}^{0} \delta E_h \cdot \frac{\partial f_{0j}}{\partial \vec{v}} \exp\left[i\left\{\vec{K}.\left(\vec{r}' - \vec{r}\right) - \Omega(t' - t)\right\}\right] dt' \\
&= \frac{e}{m} \int_{-\infty}^{0} \left(\delta E_{hy} \frac{\partial}{\partial v_y'} + \delta E_{hz} \frac{\partial}{\partial v_z'}\right) f_{0j} \times \exp\left[\imath\left\{K_{\perp}(y' - y) + K_{\|}(z - z') - \Omega\tau\right\}\right] d\tau \\
&= \frac{e}{m} \int_{-\infty}^{0} \left\{\delta E_{hy}\left(-\frac{m}{T_j} v_y' + \frac{\lambda}{\Omega_j}\right) + \delta E_{hz} \frac{\partial}{\partial v_{\|}'}\right\} f_{0j} \times \exp\left[i\left\{K_{\perp}(y' - y) + K_{\|}(z' - z) - \Omega\tau\right\}\right] d\tau \\
\delta f_h &= \frac{e}{m} f_{0j} \delta E_{hy} \int_{-\infty}^{0} \left(-\frac{m}{T_j} v_y' + \frac{\lambda}{\Omega_j}\right) \times \exp\left[i\left\{K_{\perp}(y' - y) + K_{\|}(z' - z) - \Omega\tau\right\}\right] d\tau \\
&\quad + \frac{e}{m} \delta E_{hz} \frac{\partial}{\partial v_{\|}} f_{0j} \int_{-\infty}^{0} \exp\left[i\left\{K_{\perp}(y' - y) + K_{\|}(z' - z) - \Omega\tau\right\}\right] d\tau
\end{aligned} \tag{16}$$

We can now write the expression $\delta f_h\left(\vec{K}, \Omega\right)$ in the form

$$\delta f_h(K, \Omega) = \frac{ie}{m} \frac{\delta E_h}{\left|\vec{K}\right|} \frac{m}{T_j} \left[1 + \left(\Omega - \frac{\lambda T_j K_{\perp}}{m\Omega_j}\right) \sum_a \sum_b \frac{J_a(\alpha) J_b(\alpha) \exp\left\{i(a - b)\left(\theta - \frac{\pi}{2}\right)\right\}}{K_{\|} v_{\|} - \Omega - \lambda' \frac{v_{\perp}^2 K_{\perp}}{2\Omega_j} + a\Omega_j}\right] \times f_{0j} \tag{17}$$

Next, we evaluate Eq. (12) to obtain δf_{lh} as follows.
We write

$$\delta f_{lh}\left(\vec{K},\Omega\right) = I_1^{th} + I_2^{th} + I_3^{th} \tag{18}$$

The first part is

$$\begin{aligned} I_1^{th} &= \frac{e}{m}\int_{-\infty}^{0} E_{l\parallel}\frac{\partial}{\partial v_\parallel}(\delta f_h)\exp\left[i\left\{\left(\vec{K}-\vec{k}\right).\left(\vec{r'}-\vec{r}\right)-(\Omega-\omega)\tau\right\}\right]d\tau \\ &= \frac{-ie}{m}E_{l\parallel}\frac{\partial}{\partial v_\parallel}\left[\frac{ie}{m}\frac{\delta E_h}{\left|\vec{K}\right|}\frac{m}{T_j}f_{0j}\left\{1+\sum_a\sum_b\frac{J_a(\alpha)J_b(\alpha)\exp\left\{i(a-b)\left(\theta-\frac{\pi}{2}\right)\right\}}{K_\parallel v_\parallel-\Omega-\frac{\lambda' v_\perp^2 K_\perp}{2\Omega_j}+a\Omega_j}\right.\right. \\ &\quad \left.\left.\times\left(\Omega-\frac{\lambda T_j K_\perp}{m\Omega_j}\right)\right\}\right]\times\left(\sum_s\sum_t\frac{J_s(\alpha)J_t(\alpha)\exp\left\{i(s-t)\left(\theta-\frac{\pi}{2}\right)\right\}}{(K_\parallel-k_\parallel)v_\parallel-\Omega'-\frac{\lambda' v_\perp^2 K_\perp}{2\Omega_j}+s\Omega_j}\right) \end{aligned} \tag{19}$$

The second part is

$$\begin{aligned} I_2^{lh} &= \frac{e}{m}\int_{-\infty}^{0}\left(E_{hy}\frac{\partial}{\partial v'_y}+\frac{\partial}{\partial v'_z}\right)f_{1j}\times\exp\left[i\left\{\left(\vec{K}-\vec{k}\right).\left(\vec{r}'-\vec{r}\right)-(\Omega-\omega)\tau\right\}\right]d\tau \\ &= \frac{e}{m}\int_{-\infty}^{0}\left[E_{hy}\left\{\frac{ie}{m}\frac{E_{l\parallel}}{\omega'-k'_\parallel v_\parallel+i0^+}\frac{\partial}{\partial v_\parallel}\left(\frac{\partial}{\partial v'_y}f_{0j}\right)\right\}\right. \\ &\quad \left.+E_{hz}\left\{\frac{\partial}{\partial v_\parallel}\left(\frac{ie}{m}\frac{E_{l\parallel}}{\omega'-k_\parallel v_\parallel+i0^+}\frac{\partial}{\partial v_\parallel}f_{0j}\right)\right\}\right] \\ &\quad \times\exp\left[i\left\{\left(\vec{K}-\vec{k}\right).\left(\vec{r}'-\vec{r}\right)-(\Omega-\omega)\tau\right\}\right]d\tau \\ &= \frac{e}{m}\int_{-\infty}^{0}\frac{\delta E_h}{\left|\vec{K}\right|}\left[\left\{K_\perp\frac{E_{l\parallel}}{\omega'-k_\parallel v_\parallel+i0^+}\left(\frac{\partial}{\partial v_\parallel}f_{0j}\right)\times\left(-\frac{m}{T_j}v'_y+\frac{\lambda}{\Omega_j}\right)\right\}\right. \\ &\quad \left.+\left\{K_\parallel\frac{\partial}{\partial v_\parallel}\frac{ie}{m}\frac{E_{l\parallel}}{\omega'-k_\parallel v_\parallel+i0^+}\left(\frac{\partial}{\partial v_\parallel}f_{0j}\right)\right\}\right] \\ &\quad \times\exp\left[i\left\{\left(\vec{K}-\vec{k}\right).\left(\vec{r}'-\vec{r}\right)-(\Omega-\omega)\tau\right\}\right]d\tau \end{aligned} \tag{20}$$

$$I_2^{lh} = \frac{-ie\,\delta E_h}{m\left|\vec{K}\right|}\left[-\frac{m}{T_j}\left(\frac{ie}{m}\frac{E_{l\|}}{\omega' - k'_\| v_\| + i0^+}\right)\left\{1 + \left(\Omega' - v_\| K'_\|\right)\right.\right.$$

$$\left.\times\sum_a\sum_b\frac{J_a(\alpha)J_b(\alpha)\exp\{i(a-b)\theta'\}}{a\Omega_j - \dfrac{K_\perp\lambda v_\perp^2}{2\Omega_j} + v_\| K'_\| - \Omega'}\right\}\frac{\partial}{\partial v_\|}f_{0j}$$

$$\left.+K_\|\sum_a\sum_b\frac{J_a(\alpha)J_b(\alpha)\exp\{i(s-t)\theta'\}}{s\Omega_j - \dfrac{K_\perp\lambda' v_\perp^2}{2\Omega_j} + v_\| K'_\| - \Omega'}\times\frac{\partial}{\partial v_\|}\left(\frac{ie}{m}\frac{E_{l\|}}{\omega' - k'_\| v_\| + i0^+}\right)\frac{\partial}{\partial v_\|}f_{0j}\right] \tag{21}$$

and
The third part is

$$I_3^{lh} = \frac{e}{m}\int_{-\infty}^{0}\delta\vec{E}_{lh}.\frac{\partial}{\partial\vec{v}}f_{0j}\times\exp\left[i\left\{\left(\vec{K}-\vec{k}\right).\left(\vec{r}'-\vec{r}\right) - (\Omega-\omega)\tau\right\}\right]d\tau$$

$$= \frac{e}{m}\int_{-\infty}^{0}\delta\vec{E}_{lh}\left(-\vec{v}\frac{m}{T_j} + \vec{y}\frac{\lambda}{\Omega_j}\right)f_{0j}\times\exp\left[i\left\{\left(\vec{K}-\vec{k}\right).\left(\vec{r}'-\vec{r}\right) - (\Omega-\omega)\tau\right\}\right]d\tau$$

$$I_3^{lh} = \frac{e}{m}\int_{-\infty}^{0}\frac{\delta\vec{E}_{lh}\left(\vec{K}-\vec{k}\right)}{\left|\vec{K}-\vec{k}\right|^2}\left\{-\frac{m}{T_j}\left(\vec{K}-\vec{k}\right).\vec{v} + \vec{y}\left(\vec{K}-\vec{k}\right)\frac{\lambda}{\Omega_j}\right\}f_{0j}$$

$$\times\exp\left[i\left\{\left(\vec{K}-\vec{k}\right).\left(\vec{r}'-\vec{r}\right) - (\Omega-\omega)\tau\right\}\right]d\tau$$

$$I_3^{lh} = \frac{e}{im}\frac{\delta\vec{E}_{lh}}{\left|\vec{K}-\vec{k}\right|}\left(-\frac{m}{T_j}\right)f_{0j}\int_{-\infty}^{0}\left[1 + \left(\Omega - \omega - \frac{\lambda K_\perp T_j}{\Omega_j m}\right)\right.$$

$$\left.\times\exp\left[i\left\{\left(\vec{K}-\vec{k}\right).\left(\vec{r}'-\vec{r}\right) - (\Omega-\omega)\tau\right\}\right]\right]d\tau$$

$$= \frac{ie}{m}\frac{\delta\vec{E}_{lh}}{\left|\vec{K}-\vec{k}\right|}\frac{m}{T_j}f_{0j}\left\{1 + \left(\Omega' - \frac{\lambda K_\perp T_j}{\Omega_j m}\right)\sum_{a,b}\frac{J_a(\alpha)J_b(\alpha)\exp\{i(a-b)\theta'\}}{a\Omega_j - \dfrac{K_\perp\lambda' v_\perp^2}{2\Omega_j} - v_\| K'_\|\Omega'}\right\} \tag{22}$$

Finally, the high fluctuating part is obtained as follows:
We write

$$\Delta f = I_1^{\Delta f} + I_2^{\Delta f} \tag{23}$$

The first part is evaluated as

$$\begin{aligned} I_1^{\Delta f} &= \frac{e}{m} E_{l\|} \int_{-\infty}^{0} \frac{\partial}{\partial v_\|} (\delta f_{lh}) \exp\left[i\left\{\vec{K}.\left(\vec{r}' - \vec{r}\right) - \Omega\tau\right\}\right] d\tau \\ &= \frac{-ie}{m} E_{l\|} \sum_{a,b} \frac{J_a(\alpha) J_b(\alpha) \exp\{i(a-b)\theta'\}}{a\Omega_j - \frac{K_\perp \lambda' v_\perp^2}{2\Omega_j} - \Omega} \frac{\partial}{\partial v_\|} \delta f_{lh} \end{aligned} \tag{24}$$

The second part is

$$\begin{aligned} I_2^{\Delta f} &= \frac{e}{m} \int_{-\infty}^{0} \delta \vec{E}_{lh} \cdot \frac{\partial}{\partial \vec{v}} \left(\frac{ie}{m\,\omega'' - k_\|'' v_\| + i0^+} E_{l\|} \frac{\partial}{\partial v_\|} f_{0j} \right) \exp\left[i\left\{\vec{K}.\left(\vec{r}' - \vec{r}\right) - \Omega\tau\right\}\right] d\tau \\ &= \frac{e}{m} \int_{-\infty}^{0} \left[\delta E_{lhy} \frac{\partial}{\partial v_y'} \left(\frac{ie}{m\,\omega - k_\| v_\| + i0^+} E_{l\|} \frac{\partial}{\partial v_\|} f_{0j} \right) + E_{lhz} \frac{\partial}{\partial v_\|} \left(\frac{ie}{m\,\omega - k_\| v_\| + i0^+} E_{l\|} \frac{\partial}{\partial v_\|} f_{0j} \right) \right] \times \exp\left[i\left\{\vec{K}.\left(\vec{r}' - \vec{r}\right) - \Omega\tau\right\}\right] d\tau \end{aligned} \tag{25}$$

Which takes the form

$$\begin{aligned} I_2^{\Delta f} &= \frac{e}{m} \frac{\delta E_{lh}}{\left|\vec{K} - \vec{k}\right|} \int_{-\infty}^{0} \left[K_\perp \left\{ \frac{ie}{m\,\omega - k_\| v_\| + i0^+} E_{l\|} \frac{\partial}{\partial v_\|} \left(\frac{\partial}{\partial v_y'} f_{oj} \right) \right\} + (K_\| - k_\|) \frac{\partial}{\partial v_\|} \left\{ \frac{ie}{m\,\omega - k_\| v_\| + i0^+} E_{l\|} \frac{\partial}{\partial v_\|} f_{oj} \right\} \right] \\ &\quad \times \exp\left[i\left\{\vec{K}.\left(\vec{r}' - \vec{r}\right) - \Omega\tau\right\}\right] d\tau \\ &= \frac{-e}{m} \frac{\delta E_{lh}}{\left|\vec{K} - \vec{k}\right|} \left[\left(\frac{ie}{m\,\omega - k_\| v_\| + i0^+} E_{l\|} \frac{\partial}{\partial v_\|} f_{oj} \right) \left[1 + \left(\Omega - K_\| v_\| - \frac{\lambda T_j}{m\Omega_j} \right) \times \sum_{a,b} \frac{J_s(\alpha) J_t(\alpha) \exp\{i(s-t)\theta'\}}{s\Omega_j - \frac{K_\perp \lambda' v_\perp^2}{2\Omega_j} + K_\| v_\| - \Omega} \right] + \right. \\ &\quad \left. (K_\| - k_\|) \frac{ie}{m} \frac{\partial}{\partial v_\|} \times \left(\frac{E_{l\|}}{\omega - k_\| v_\| + i0^+} \frac{\partial}{\partial v_\|} f_{oj} \right) \sum_{a,b} \frac{J_s(\alpha) J_t(\alpha) \exp\{i(s-t)\theta'\}}{s\Omega_j - \frac{K_\perp \lambda' v_\perp^2}{2\Omega_j} + K_\| v_\| - \Omega} \right] \end{aligned}$$

We can thus write the expression $\Delta f(K, \Omega)$ in the form of

$$\Delta f = \frac{-ie}{m} E_{l\parallel} \sum_a \sum_b \frac{J_a(\alpha) J_b(\alpha) \exp\{i(a-b)\theta'\}}{a\Omega_j - \frac{K_\perp \lambda' v_\perp^2}{2\Omega_j} - \Omega} \frac{\partial}{\partial v_\parallel} \delta f_{lh} + \frac{-e}{m} \frac{\delta E_{lh}}{\left|\vec{K} - \vec{k}\right|} \left[\left(\frac{ie}{m} \frac{E_{l\parallel}}{\omega - k_\parallel v_\parallel + i0^+} \frac{\partial}{\partial v_\parallel} f_{0j} \right) \right.$$

$$\left[1 + \left(\Omega - K_\parallel v_\parallel - \frac{\lambda T_j}{\Omega_j m} \right) \times \sum_{s,t} \frac{J_s(\alpha) J_t(\alpha) \exp\{i(s-t)\theta'\}}{s\Omega_j - \frac{K_\perp \lambda' v_\perp^2}{2\Omega_j} + K_\parallel v_\parallel - \Omega} \right] +$$

$$\left. K'_\parallel \frac{ie}{m} \frac{\partial}{\partial v_\parallel} \times \left(\frac{E_{l\parallel}}{\omega - k_\parallel v_\parallel + i0^+} \frac{\partial}{\partial v_\parallel} f_{0j} \right) \sum_{s,t} \frac{J_s(\alpha) J_t(\alpha) \exp\{i(s-t)\theta'\}}{s\Omega_j - \frac{K_\perp \lambda' v_\perp^2}{2\Omega_j} + K_\parallel v_\parallel - \Omega} \right] \tag{26}$$

where $\Omega' = \Omega - \omega, K'_\parallel = K_\parallel - k_\parallel$.

The nonlinear dielectric function of the test high frequency Bernstein wave in the presence of drift wave turbulence is then calculated using the Vlasov Poisson system of equations.

$$\nabla.\delta\vec{E}_h = -4\pi jen_j \int \left(\delta f_{hj} + \Delta f_j \right) dv \tag{27}$$

which becomes

$$\delta E_h \left(\vec{K}, \Omega \right) \varepsilon_h \left(\vec{K}, \Omega \right) = 0 \tag{28}$$

We write the dispersion relation $\varepsilon_h \left(\vec{K}, \Omega \right)$ in the form

$$\varepsilon_h \left(\vec{K}, \Omega \right) - \varepsilon_0 \left(\vec{K}, \Omega \right) + \varepsilon_d \left(\vec{K}, \Omega \right) + \varepsilon_p \left(\vec{K}, \Omega \right) \tag{29}$$

Where $\varepsilon_0 \left(\vec{K}, \Omega \right)$ is the linear part, $\varepsilon_d \left(K, \Omega \right)$ is the direct coupling part and $\varepsilon_p \left(\vec{K}, \Omega \right)$ is the polarisation coupling part.

4. Conclusions

In this chapter, we want to propose a nonlinear dispersion relation of electrostatic Bernstein mode waves that is established in the presence of drift wave turbulence, which is a common phenomenon in an inhomogeneous plasma. The fluctuating parts f_{1j}, δf_h, δf_{lh} and, Δf, as outlined in Eqs. (15), (17), (18), and (26) respectively, have been obtained. The fluctuating part f_{1j} is a result of the drift wave turbulent field, which is characterised by its linear nature. Furthermore, δf_h denotes the fluctuating part of the particle distribution function resulting from the perturbed electrostatic Bernstein mode, whereas δf_{lh} and Δf represents the nonlinear fluctuating parts of the distribution function. The expressions of Δf includes polarisation coupling and direct coupling terms. The nonlinear dispersion relation will be given by

$\varepsilon_h\left(\vec{K},\Omega\right) = \varepsilon_0\left(\vec{K},\Omega\right) + \varepsilon_d\left(\vec{K},\Omega\right) + \varepsilon_p\left(\vec{K},\Omega\right)$ which includes the linear part $\varepsilon_0\left(\vec{K},\Omega\right)$, the direct coupling part $\varepsilon_d\left(\vec{K},\Omega\right)$, and the polarisation coupling part $\varepsilon_p\left(\vec{K},\Omega\right)$.

5. Discussion and conclusion

The expression of the nonlinear dispersion relation involves gradient parameters associated with density and temperature gradients, along with the external force term. The growth of Bernstein mode can be predicted from this nonlinear dispersion relation. It is clearly observed that the growth of Bernstein mode is connected with drift wave turbulent field energy, and this growth may be influenced by the parameter λ associated with the density gradient and temperature gradient.

In this present investigation, the external force field F is made to be involved in the dispersion relation and in the growth of Bernstein mode. In all previous investigations on the growth of Bernstein mode, the external force field F was not considered.

Conflict of interest

The authors declare no conflict of interest.

Author details

Banashree Saikia* and Paramananda Deka
Department of Mathematics, Dibrugarh University, Dibrugarh, Assam, India

*Address all correspondence to: banashrees13@gmail.com

IntechOpen

© 2024 The Author(s). Licensee IntechOpen. This chapter is distributed under the terms of the Creative Commons Attribution License (http://creativecommons.org/licenses/by/3.0), which permits unrestricted use, distribution, and reproduction in any medium, provided the original work is properly cited.

References

[1] Stix TH. Waves in Plasmas. Springer Science & Business Media; 1992

[2] Kurth WS, Craven JD, Frank LA, Gurnett DA. Journal of Geophysical Research. 1979;**84**:4145-4164. DOI: 10.1029/JA084iA08

[3] Grimald S, Decreau PME, Canu P, Rochel A, Vallies X. Journal of Geophysical Research. 2008;**113**:A11216. DOI: 10.1029/2008JA013290

[4] Deka PN, Borgohain A. On unstable electromagnetic radiation through nonlinear wave–particle interactions in presence of drift wave turbulence. Journal of Plasma Physics. 2012;**78**(5): 515-524

[5] Senapati P, Deka PN. Instability of electron bernstein mode in presence of drift wave turbulence associated with density and temperature gradients. Journal of Fusion Energy. 2020;**39**(6): 477-490

[6] Singh M, Deka PN. Plasma-maser effect in inhomogeneous plasma in the presence of drift wave turbulence. Physics of Plasmas. 2005;**12**(10)

[7] Saikia B, Deka PN. Generation of O-mode in the presence of ion-cyclotron drift wave turbulence in a nonuniform plasma. East European Journal of Physics. 2023;(3):122-132

[8] Singh M. Plasma-maser instability of the electromagnetic radiation. In: The Presence of Electrostatic Drift Wave Turbulence in Inhomogeneous Plasma

[9] Deka PN. Orthogonal interaction of Bernstein mode with ion-acoustic wave through plasma maser effect. Pramana. 1998;**50**:345-354

[10] Singh M. Study of nonlinear interaction of waves through plasma-maser in magnetosphere plasma. Planetary and Space Science. 2007;**55**(4): 467-474

[11] Singh M, Deka PN. Plasma-maser instability of the ion acoustics wave in the presence of lower hybrid wave turbulence in inhomogeneous plasma. Pramana. 2006;**66**:547-561

[12] Yoon PH, Hadi F, Qamar A. Bernstein instability driven by thermal ring distribution. Physics of Plasmas. 2014;**21**(7)

[13] Henning FD, Mace RL, Pillay SR. Electrostatic Bernstein waves in plasmas whose electrons have a dual kappa distribution: Applications to the Saturnian magnetosphere. Journal of Geophysical Research: Space Physics. 2011;**116**(A12)

[14] Nambu M. Plasma-maser effects in 62 plasma astrophysics. Space Science 63 Reviews. 1986;**44**(3):357-391

[15] Nambu M, Saikia BJ, Gyobu D, Sakai JI. Nonlinear plasma maser driven by electron beam instability. Physics of Plasmas. 1999;**6**(3):994-1002

[16] Nambu M. Plasma-maser effects in plasma astrophysics. Space Science Reviews. 1986;**44**(3-4):357-391

[17] Sarma SN, Sarma KK, Nambu M. Plasma maser theory of the extraordinary mode in the presence of Langmuir turbulence. Journal of Plasma Physics. 1991;**46**(2):331-346

[18] Ichimaru S. Statistical Plasma Physics, Volume I: Basic Principles. CRC Press; 2018